The Magic of Living Consciousness

Eugene Subbotsky

The Magic of Living Consciousness

The Wonders of the Mundane

palgrave
macmillan

Eugene Subbotsky
Department of Psychology
Lancaster University
Lancaster, UK

ISBN 978-3-031-54707-2 ISBN 978-3-031-54708-9 (eBook)
https://doi.org/10.1007/978-3-031-54708-9

Cover credit: Thinkstock Images/gettyimages

This Palgrave Macmillan imprint is published by the registered company Springer Nature Switzerland AG
The registered company address is: Gewerbestrasse 11, 6330 Cham, Switzerland

Paper in this product is recyclable.

Contents

List of Figures

List of Tables

PART I

Living Consciousness in Communication and Education

CHAPTER 1

Living Consciousness: Introduction

1.1 Living Consciousness

Fascinated by modern science, we are doing our best to be in harmony with the rules of reason, but it doesn't always work out. Freud was one of the first psychologists who noticed this and claimed for a person the right to be irrational. Still, he didn't go far enough, and placed the irrational in the domain of unconscious.

But take a closer look at what's going on in our conscious mind when it is hidden from the eyes and ears of others, and you will see that our thoughts and feelings, like Adam and Eve in the Garden of Eden, are blissfully unaware of logic and morality. Yet the fall of man is unavoidable. Like Adam and Eve after eating from the prohibited tree, our secret thoughts and feelings lose their magical freedom as soon as they turn into knowledge shared with others. We become rational individuals when our private consciousness solidifies into spoken or written language, logical thinking, and moral behavior. But crossing the edge of rationality comes at a cost: To embrace logic and morality, we have to part with the magic of creativity. In this book, we will explore our private subjective world in detail.

Sometimes we are so used to wonderful things and events that we stop noticing them. As if enchanted by a magic spell, these things become invisible. Our aim is to de-spell the enchanted things and show that what we do every day without paying attention to it is in fact full of

E. Subbotsky, *The Magic of Living Consciousness*,
https://doi.org/10.1007/978-3-031-54708-9_1

mysteries. The enchanted thing I have in mind is our inner subjective reality. We will examine the laws that govern our inner mental processes (creative thinking, imagining, dreaming, and feeling), and discuss how these processes can be observed, studied, and used for practical purposes in life. So, let's begin.

Imagine that on a beautiful sunny day you are strolling in a park. At one moment you feel tired and sit on the bench under a tree. You close your eyes and sit back, trying to relax. Suddenly you realize that despite you are completely motionless and alone, some work is going on in your mind. You remember an episode from your childhood, then, by association, return to the present, next you think about some events in the future. If you are a scientist, you keep thinking about the problem you have been trying to crack in the nearest few days, if you are an artist—searching for an image best suitable for a painting or a novel, if you are.... To cut a long story short, despite abstaining from external activity and direct communication with others, inside your mind you keep being actively involved into problem solving, memorizing, imagining, and wanting something. Some of these mental processes you can directly control with your internal Self, others, like fears or night dreams, appear and disappear independently of your purposeful efforts. This private activity of your mind, hidden from others, is your *conscious living consciousness*. It is *private* because no one else has access to it unless you let it out by speaking loudly or acting with your body, *conscious* because you are aware of it, *living* because it is happening here and now, and *consciousness* because it unfolds in the form of subjective experiences.

When you open your eyes, you get into the world around you, which is full of physical objects. You see the trees of the park, accurately tended lawns, and alleys. There, behind the trees of the park, you see silhouettes of high buildings and hear the sound of moving cars. All of this is created by people with the help of their living consciousness, and so it belongs to consciousness too, but now it has become 'dead', *objectified consciousness*, which we usually call matter. Unlike living consciousness, objectified consciousness is accessible to everyone. Much of the objectified consciousness, like cars and buildings, is a creation of people's living consciousness, another part, like mountains, trees, and birds, is a creation of God or Nature, but the common feature of all the things of objectified consciousness is that they exists objectively, outside your mind, and are independent of your internal Self. When you open a handbook on physics, which till this moment has been stored in your briefcase, you see letters, numbers, characters, logical statements, and formulas. Letters, formulas, concepts, and logical thinking belong to objectified consciousness as well.

Next, you notice that living consciousness and objectified consciousness are separate but not isolated parts of reality. There is a link or exchange between them. By looking at the building or the formulas, by hearing a lecture or a song, you take the objectified consciousness inside your mind, or appropriate it, by making it a part of your living consciousness, in the form of inner images or thoughts. When objectified consciousness becomes a part of your living consciousness, it loses its permanence and locality. Indeed, when thinking about a certain building that you have in front of you, you know that it has a rectangular shape and a size that can be measured in meters and doesn't change unless the building is reconstructed or destroyed. By taking the building inside your living consciousness, you can mentally make it higher or lower, and change its shape, size, and location. When you learn about the laws of nature or logical thinking, you can start 'playing' with them by imagining that 2 plus 2 equals 5, or that your body can fly in the sky denying the law of gravity. In other words, your creative internal Self can experiment with the appropriated objectified consciousness. Usually, we call this process the work of the imagination. If you are an artist, you might wish to return the results of this shake-up in the form of a painting like those by Rene Magritte or Salvador Dali, or a novel like those by Franz Kafka or Mikhail Bulgakov, but if you are a scientist then much of your creative combinations might go to trash, yet some might result in a new and ground shaking theory, like special relativity theory by Albert Einstein.

Sometimes it seems to us that we think in words. L. S. Vygotsky wrote: "Thought is accomplished in word, and not only expressed in it" [1, p. 162]. But words is an instance of objectified consciousness. The word 'bird' has nothing in common with the feathered two-legged creature with wings. When a baby is born, the baby doesn't have language, but he or she can think through comparing images with each other and making appropriate conclusions. For example, even before birth the baby can distinguish the voice of his or her mother from the voice of a stranger, a newborn baby can distinguish between a picture of a scrambled face and a picture of a regular face, and at the age of three and a half months children show surprise when a large object moving from left to right disappears behind an opaque screen and appears on the other side of the screen without flashing through the window cut out in the middle of the screen. Finally, it is not rare that we have a thought but don't have a word to express the thought. A poet searches for the right words to 'dress' his or her feelings into the socially acceptable 'cloths', and a scientist thinks not in words but in images that are still in need of being 'dressed in flesh' of words or numbers.

Although the irrational unconscious processes in the form of hypnotic states have been known since medieval times [2, 3], it was Sigmund Freud who gave irrationality a legitimate place in people's minds, by placing it into the domain of unconscious [4]. There, in the unconscious 'Id' the innate primaeval drives operate in accord with the 'pleasure principle', supplying the conscious 'Ego' with energy. It is in the 'Id' where our 'slips of the tongue' and neurotic complexes are born, when the irrational and immoral drives meet with the 'reality principle' in the form of the laws of logic and morality. But is the unconscious mind the only domain where the irrational processes reign? Having pondered the way our living consciousness works we see that it isn't.

Indeed, when we look at our living consciousness, we see that irrationality flourishes there as well. Thoughts and images appear from nothing and disappear into thin air, images merge together to form the most strange and bizarre combinations, in a dream people can transform into animals or inanimate objects, we can travel back in time, go through solid walls, talk with animals, and do things that in the real life would be qualified as criminal or immoral. The same magical things happen to our feelings. We see a person or a piece of art and feel that we don't like them, even though we don't know anything about the person and never seen this piece of art before. Conversely, we can see another person and a piece of art and instantly fall in love with them. Psychologists call this ability of living consciousness 'participation'.

When we are dining on a veranda of a nice restaurant and see a dead bird in the garden or feel the smell of rotten flesh, this can spoil the taste of the dish we are eating, even if the dish was cooked by an exquisite chef. When shaking hands with a terminally ill person, we might have an unpleasant feeling that the illness may stick to us despite knowing that the illness is not contagious. But the most prominent manifestation of living consciousness is faith. Faith is the ability of our internal Self to identify with something we can't perceive but can imagine. Even if we don't believe in God, we still believe in a lot of things. We believe that the world will keep on existing after our death, even though we will never see it. We believe that a feather put in a vacuum will always fall on Earth with acceleration of 9.8 m/sec^2, even though in order to be certain of that we would have to make an unlimited number of measurements, which is impossible. We believe that two parallel lines on a flat surface will never meet, even though we are unable to personally trace the lines into infinity. When objectified, our faith turns into the 'unbreakable' laws of nature, mathematics and logic. Finally, living consciousness contains the most precious gem of the mind—our *internal Self*. It is our internal Self that gives us the sense of freedom—the ability to choose without compulsion. In other words, in our inner subjective world almost all things behave like in a magical dream.

Almost, but not all. Some of the objectified consciousness manages to get inside our mind. Mental calculation, logical thinking, rational planning inside our mind resists the pressure of magic, despite such resistance requires a considerable effort from our internal Self. It is easy to calculate with the help of a calculator, but mental calculation entails a substantial probability of making an error. It doesn't take much effort to memorize a telephone number by writing it down but keeping the number in mind without the help of writing is much harder. When we learn foreign languages, we might struggle to remember the unfamiliar words without using memory aids. When we are criticized by our boss, we might have negative feelings toward him or her but behave as if we like the person. In other words, inside our mind there is a bridge that connects our living consciousness to objectified consciousness and submits our private feelings to moral and social norms. The objectified consciousness that functions inside our mind is the *internal objectified consciousness*. The internal objectified consciousness is a buffer domain that links our living consciousness with external objectified consciousness. Like our living consciousness, internal objectified consciousness is private and unfolds 'here and now' inside our mind. Unlike our living consciousness, the internal objectified consciousness conforms to the laws of formal logic and rational thinking. But most importantly, our living consciousness and internal objectified consciousness have different roles within our mind. Whereas our living consciousness is a generator of creativity, the internal objectified consciousness is a convertor.

More specifically, a new idea results from the creative impulse in the domain of living consciousness, and then may take two different routes. The idea takes the *rout of rational conversion* when it enters the *buffer zone* of the internal objectified consciousness where it is converted into logical and rational forms with the help of inner speech, and then proceeds into the domain of external objectified consciousness, where it takes the form of a literary work, a scientific discovery, realistic art or rational behaviour. In contrast, the idea follows the ***rout of direct intrusion*** when it gets into the domain of external objectified consciousness bypassing the buffer zone; as a result, we get a 'crazy' idea, a piece of magical art or a magical behaviour (see Fig. 1.1).

Whereas the internal objectified consciousness is studied by such disciplines as logic, psycholinguistics and linguistics, and the external objectified consciousness is a favourite subject of specialists in artificial

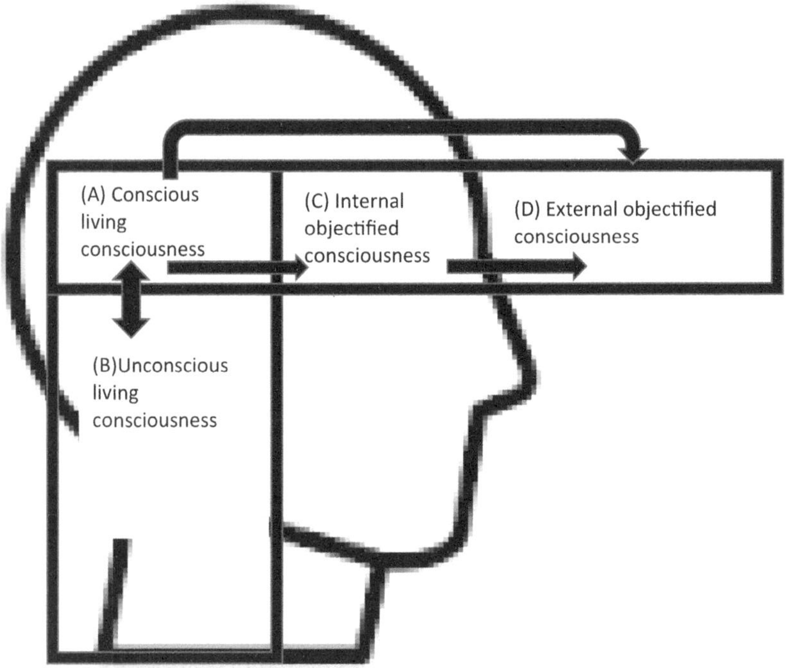

Fig. 1.1 **a** Conscious living consciousness (private subjective reality that functions in accord with the laws of magic), **b** Unconscious living consciousness (unconscious subjective reality that functions in accord with the laws of magic), **c** Internal objectified consciousness (private subjective reality that functions in accord with the laws of formal logic and rationality), and **d** External objectified consciousness (symbolically presented laws of logic and science, verbal and actual behaviour and human artifacts). The route of rational conversion **a** to **c** to **d**. The route of direct intrusion **a** to **d**

intelligence, languages and natural sciences, the living consciousness has been given much lesser attention, and for obvious reasons. One reason is that we study objectified consciousness with scientific methods, from the 3rd person perspective. Physical objects (e.g., minerals, plants, and animals) are first perceived as phenomena, and then compared, measured, and turned into symbols: numbers, blueprints, names, and notions. The same happens to human artifacts (e.g., man-made objects, languages, and concepts): we perceive them as external objects and then remember their meanings and operate with them. This is an easy task since objectified consciousness conforms to the laws of permanence and locality: it can be fixed in the form of external things, words or symbols, and operated with at leisure. But how can we approach living consciousness, when our own 'observing device'—our internal Self—is a part of our living consciousness and as such is invisible for itself? Besides, out thoughts, images and memories are in the process of permanent change, conversion, and transformation. This is one reason why living consciousness has been mostly studied introspectively by philosophers, artists, and writers.

Another reason of underappreciation of living consciousness is the magnificent success of science in the modern world, which has plunged many scientists into the illusion that magical events are ancient history and exist today only in play, art, and night dreams. The illusion reached its pinnacle in the middle of the XXth century, when semiconductors, portable radios, television, nuclear power stations, flights to the Moon, early computers, genetic engineering, and other wonders of science made some scientists believe that there is nothings in the world that cannot be explained by science. But there was a price to pay for this scientific optimism—the scientists became blind to their living consciousness. They began looking at the world as if nature and objectified consciousness were the only things that mattered, with living consciousness being viewed as an illusion that accompanies brain processes but has no causal powers. Inspired by the achievements of science, physicists began searching for the 'theory of everything'—a certain unified set of laws from which it would be possible to deduct all the existing entities and processes. Scientists specializing in the AI began dreaming of creating the digital copy of human consciousness. But physics study inanimate things and not people, the AI lacks the core element of living consciousness—life, and the would be 'theory of everything' lacks the most essential part of the 'everything'—the living consciousness of physicists. The living consciousness fell 'under the spell' of science and cybernetics. Big mistake!

Consequently, the aim of this book is to de-spell living consciousness and show that living consciousness is the domain of consciousness in which we spend most of our waking time and a significant part of our sleeping time. When we are looking at paintings in a museum, listening to music in a theatre, watching a movie, seeing a night dream, scrolling through our i-phone, or just wondering inside our mind—we are in the domain of the living consciousness. But most important, living consciousness is the source of the objectified consciousness. We will examine how our living consciousness functions, and how our understanding of this functioning helps in solving some key problems of modern life, such as facilitating creativity, protecting from magical manipulation of the minds, fighting certain kinds of crime, managing fake-reality, preventing misuses in psychotherapy, tackling controversial issues in science, tracing origins of totalitarian media narratives that trigger hatred and wars, and helping a person to cope with the demands of society.

1.2 What is Covered

Part I of this book 'Living Consciousness in Communication and Education' is given to examining magical thinking in the domain of learning and human interaction. In Chapter 1'Living consciousness: Introduction' The distinction is proposed between living consciousness, which includes subjective experiences not intended for sharing with others and functioning in accord with the laws of magic, versus objectified consciousness, which is living consciousness converted into forms prepared for the consumption by society and culture (e.g., scientific concepts, logical thinking, moral norms, and human artifacts) and conforms to the laws of nature and formal logic. The aim of the book is outlined as examining properties of living consciousness and showing that living consciousness, far from being a mere accompaniment to brain functions, is the primary source from which objectified consciousness is built. New points that are put forward are highlighted and a prospective audience is outlined.

In Chapter 2 'Living Consciousness and the Cultural-historical Method: Limits of Applicability' the methods of studying living consciousness and objectified consciousness are discussed. The cultural-historical method aims at managing objectified consciousness in the form of knowledge and skills with the help of symbolic structures, such as speech, logical thinking, and scientific concepts. The cultural-historical method is good at teaching readymade knowledge in school, developing skills, restoring lost cognitive functions in patients, or forming cognitive and socio-moral forms of behaviour based on external control. However, this method is inapplicable to studying and managing living consciousness, such as feelings, emotions, imagination, genuine morality, and creative thinking.

In Chapter 3 'Magical Thinking as a Dimension of Living Consciousness' the distinction is made between cognitive and emotional dimensions of living consciousness. Unlike magical thinking, which remains a conscious practice throughout the lifespan, the belief in magic in educated adults becomes mostly unconscious. The argument is presented that magical experience and the belief in magic generate psychic energy. The chapter links together phenomena that thus far have been studied separately: magical thinking in children and adults, religious beliefs, manipulation of the minds in politics, and political terror. A special emphasis is given to the ways the energy of living consciousness can be used in education.

In Part 2 of the book 'Living Consciousness and the Universe', the role of living consciousness in creating scientific view of the universe is highlighted. In Chapter 4 'Man in the Mirror of Nature: Science and Living Consciousness' the argument is presented that Nicolaus Copernicus was right about the planets, but not about a person, and the person is indeed a centre of the universe. This assumption explains the otherwise inexplicable magical components in the picture of physical universe, such as the fine tuning of fundamental cosmological constants and the role of measurement in the 'wave function collapse'. The chapter argues that scientists build their knowledge by projecting properties of their living consciousness onto the physical and social universes. A similar projection takes place in biology. The more scientists study animal behaviour, the more they see it as a reflection of their living consciousness. The point is made that the role of science in human life, for all its importance, is auxiliary, epistemological, whereas the role of living consciousness is basic, ontological.

Chapter 5 'Is Living Consciousness a Simulation? Children's and Adults' Reactions to the "Person is a simulation" Hypothesis' is an extension of the problems raised in Chapter 4. To exclude the unavoidable magical elements in explaining the origins of living consciousness some scientists suggested that a person is a computer simulation run by some kind of Supreme Programmer. Psychological experiments showed that the simulation hypothesis met strong resistance in both children and adults. The participants were happy to imagine a Supreme Programmer who could create people like them but strongly denied the existence of such a Supreme Programmer outside their imagination. Further analysis revealed that the cause of this denial was the participants' disbelief in the possibility for the imagined object to transcend their living consciousness and enter the external reality.

Chapter 6 'Living Consciousness in Historical Context' is about the main changes in the ideas about living consciousness as they happen during the time of modernity. Until the twentieth century, European culture lived on basic ideas about living consciousness laid down by the Enlightenment. Among others, such ideas included the belief that (1) living consciousness reflects reality (Realism), (2) symbols are the repository of the objectified consciousness (Symbolism), and (3) living consciousness has a measurable degree of free will (Limited freedom of living consciousness). Approximately in the middle of the twentieth century, the basic ideas underwent a process of bifurcation, during which alternative ideas sprout from the basic ideas, and the basic ideas were strengthened and sharpened. The advantages and drawbacks of both pathways for modern humans are discussed.

In Part 3 of the book 'Living Consciousness in Modern Life and Society' magical thinking is placed on the global scale, as it functions within a society. Chapter 7 'Reason in the Cobweb: Magical Manipulation of Living Consciousness' addresses the widely spread malevolent manipulation of living consciousness that targets people's implicit belief in magic. Many such crimes are committed through personal contact of fraudsters with their victims and use suggestion rather than threat. The chapter argues that the suggestion targets not the victims' rational objectified consciousness but their living consciousness in the form of the implicit belief in magic. The concept of magical beliefs is considered and illustrated by psychological research. Types of magical manipulation of living consciousness in the domains of medicine, psychotherapy, religion, military and political terror and others are analysed. Actions that could help people to protect themselves from magical manipulation are suggested.

Chapter 8 'Fake-reality: Cheating on living consciousness on a mass scale' argues that the creation of fake-reality exploits the participatory nature of living consciousness, when a person inadvertently identifies himself or herself with a message and only a degree of the identification distinguishes rational understanding from suggestibility. Types of fake-reality in the domains of health protection, industry, cosmetics, information, politics, and others are discussed. The psychological roots of the ability to create fake messages in child development are exposed. The dynamics of fake-reality in the course of the history of European cultures is traced in art, industry, and other domains of mass culture. The distinction between good and bad fakes is introduced, and the ways of coping with fakes are discussed.

In Chapter 9, 'Totalitarian media narratives: Living consciousness in captivity' the argument is presented that one of the manifestations of living consciousness is the belief in the divine origins of political power (BDOPP). Captivated by their BDOPP and contrary to their private interests, people succumb to authoritarian messages. The link between BDOPP and two forms of political government—democracy and authoritarianism—is made. With the advent of mass media a new form of authoritarianism emerges—the totalitarian media narrative (TMN). The TMN has its psychological roots in BDOPP. Properties of TMN are analysed and causes of the emergence of competing TMNs in society are exposed, using the war in Ukraine as an example.

Chapter 10 'Golem: Living consciousness without brain' argues that a special form of living consciousness is the living consciousness of a society. Despite a society doesn't have brain of its own, it can nevertheless be represented as some kind of 'subject', which in this chapter is referred to as 'golem'. It is suggested that the most important difference that distinguishes the objectified consciousness of a golem from that of an individual is that golem's objectified consciousness is immune to the imperatives of individual morality. Ways for the individual to cope with the contradiction between the individual's objectified consciousness and that of his or her society are suggested.

In the concluding part of the book 'Mysteries of the Mundane: Epilogue and Summary' mysteries of the everyday life are summarized as phenomena of our living consciousness that escape the claws of logic and rationality.

1.3 Points Put Forward

- Living consciousness, defined as subjective experiences not addressed to the external world and conforming to the laws of magic, is opposed to objectified consciousness, which is consciousness prepared for the consumption by culture and society and conforming to the laws of logic and morality.

- Limitations of the cultural-historical method of manipulation with consciousness are outlined. This method is effective in working with objectified consciousness (e.g., teaching ready-made knowledge at school, developing skills, and restoring lost cognitive functions in patients), but is inapplicable to working with living consciousness.
- Principles of studying living consciousness are discussed, and dangers of practical work with living consciousness are exposed, such as the transformation of positively intended psychological help into self-gratification of a therapist, suspension, an overdose of influence, energy vampirism, and privacy violation.
- The widely spread fraud, which in the context of information security is known as 'social engineering', is a malevolent manipulation of people's living consciousness that exploits the tendency of living consciousness to identify with authoritative messages. Rules that help a person to protect himself or herself from magical manipulation of their living consciousness are suggested.
- It is argued that the cognitive dimension of living consciousness in the form of magical thinking and belief in magic in modern people is a source of psychological energy, which can be used for practical purposes, such as promoting creativity, improving memory, increasing motivation in learning and manipulation of the minds.
- It is argued that a scientist can explore the physical universe only through the prism of his or her private universe, by projecting properties of his or her living consciousness onto the physical universe. This assumption explains the presence of magical components in the picture of the physical universe, such as the fine tuning of fundamental physical constants, and the role of consciousness in the 'wave function collapse'.
- The book argues that the role of science in human life is auxiliary, and epistemological; in contrast, the role of living consciousness is basic, and ontological. The products of science and culture represent objectified products of living consciousness common to all people; in contrast, living consciousness is part of our personal universe that makes the basis for constructing physical and social universes.

- The hypothesis that the person is a computer simulation is interpreted as a scientistic attempt to exclude the unavoidable magical element in explaining the origins of living consciousness. Psychological experiments are presented showing that a suggestion interpreting a person as a simulation met a strong resistance in both children and educated adults, due to their disbelief that living consciousness can instantly obtain a physical embodiment.
- Changes in understanding living consciousness since the time of Renaissance till our days are traced, and the role these changes play in shaping modern life are exposed and discussed.
- It is argued that fake-reality in mass media is exploiting the property of living consciousness to inadvertently invest symbolic messages with reality.
- The point is made that the extreme manifestation of magical participation of living consciousness with social reality is people's identification with totalitarian media narratives (TMN). The reasons for the emergence of competing TMNs in society are exposed.
- It is argued that the fundamental difference between the objectified consciousness of a society and that of an individual is that the former is immune to imperatives of individual morality. Rules that could help an individual to cope with this difference are suggested.

1.4 Who Might Benefit from This Book

Undergraduates and graduates in the departments of education, politics and political psychology, philosophers and psychologists working in the area of international relations. Natural scientists may need this knowledge as well, since some natural phenomena, such as Big Bang, alternative universes, quantum entanglement and dark matter, are indistinguishable from magic. Specialists on influencing mass consciousness in politics, commercial advertising and entertainment might find this book useful as well. Educationalists may find useful the knowledge about the role of living consciousness in producing motivation for learning. Couches and psychotherapists might be interested in learning about dangers that lurk in communication between a client and a therapist. Specialists in law enforcement and information security might appreciate knowledge about the psychological origins of crimes such as 'social engineering'.

References

1 Vygotsky, L. S. (1982). The problem of consciousness (In Russ). Collected works (6 Vol., Vol. 1). Pedagogy.
2 Haque, A. (2004). Psychology from Islamic perspective: Contributions of early Muslim scholars and challenges to contemporary Muslim psychologists. *Journal of Religion and Health, 43*(4), 357–377 [365]. https://doi.org/10.1007/s10943-004-4302-z; S2CID 38740431
3 Braid, J. (2008). The discovery of hypnosis: The complete writings of James Braid, the father of Hypnotherapy. UKCHH Ltd. p. 79. ISBN 978–0–9560570–0–6.
4 Freud, S. (2013). *A general introduction to psychoanalysis*. CreateSpace Independent Publ.

CHAPTER 2

Living Consciousness and the Cultural-Historical Method: Limits of Applicability

2.1 Problem: Psyche and Living Consciousness

As stated in Chapter 1, by the term 'living consciousness' in this book we understand current and immediate subjective reality, which conforms to the laws of magic and is not exposed to the external world (see Fig. 1.1). Definitions of the term 'psyche' vary, with the most general definition linking psyche with anything that lives: "Psyche comes from the Greek psykhe, which means 'the soul, mind, spirit, or invisible animated entity which occupies the physical body'" [1]. Thus understood, psyche is an extended version of living consciousness.

But if life can be studied only from the perspective of the 3rd person, observing the behaviour and structure of living organisms, psyche can also be studied from the perspective of the 1st person, that is, through self-observation. As a result, psychology developed two approaches to the study of the psyche: subjective, through self-observation, and objective, through observation of the behaviour of living organisms. In Russia, these approaches have developed as the neo-Kantian philosophical analysis of psyche through self-observation (L.M. Lopatin, P.E. Astafiev, N.Ya. Grot, G.I. Chelpanov) and reflexological, akin to Western behaviourism (I. P. Pavlov, V. M. Bekhterev, I. M. Sechenov, K. N. Kornilov).

In psychology, there have been various attempts to affect human consciousness 'objectively', through influencing verbal or actual behaviour. One such methods, popular in modern psychology, is the

E. Subbotsky, *The Magic of Living Consciousness*,
https://doi.org/10.1007/978-3-031-54708-9_2

cultural-historical method. In this chapter, we will consider cultural-historical approach and compare this approach with other attempts to study and influence living consciousness.

In the 1930s, a group of young Russian psychologists (L. S. Vygotsky, A. R. Luria, and others) wanted to find a scientific method to study and influence living consciousness that could overcome limitations of both neo-Kantianism and behaviourism. The new method should have provided researchers with tools that allow manipulation of consciousness in the same way physicists manipulate matter. They found these tools in culture—social norms, languages, and concepts.

2.2 Living Consciousness as a Source of Higher Mental Functions

L. S. Vygotsky hypothesized that at about the age of 2 years, a child's pre-speech thinking is merged with pre-intellectual language (lower mental functions) and verbal thinking is born (a higher mental function) [2]. Considering in detail the role of speech in behaviour, L. S. Vygotsky sees this role in the person mastering his or her own behaviour, reflecting on his or her actions 'as if from the outside,' planning his or her actions, and solving problems by creating 'second-order stimuli,' such as remembering through associations or thinking. At the same time, the appearance of inner speech L. S. Vygotsky explains using the descriptive term 'interiorization,' the psychological mechanism of which remains mysterious. Indeed, why is it that a child's native and even foreign language is easily acquired (interiorized) and why is it so difficult for an adult to learn a foreign language? It would seem that the more developed the higher psychological functions are, the easier the interiorization of the external into the internal should occur, but the situation is just the opposite.

Often the hypothesis of L. S. Vygotsky that verbal thinking plays a leading role in the behaviour of an adult is accepted as an axiom. But let us ask ourselves to what extent and in what situations does speech really determine our conscious actions?

More than half a century ago, the author of this book, under the leadership and initiative of A. R. Luria, was involved in research into the so-called 'programmed behaviour' in children. We have found that children begin to subordinate their actions to verbal instructions—both those given by adults and those given by the children to themselves—at approximately the age of 2.5 years [3–5]. These data confirm the hypothesis of L. S. Vygotsky that speech does not begin to influence behaviour immediately, but only at the age of about 2 years. However, both before and after this age, most of a person's behaviour is regulated not by verbal instructions, but by living consciousness, and it is 'programmed' not by speech, but by the surrounding objective environment. Our walking, running, manual and other movements are conscious, and voluntary, but not programmed by verbal instructions, and this gives our body the opportunity to feel natural. Let us imagine, for example, that we were asked to sit on a chair, subordinating our movement not to a living consciousness, but to verbal instructions regulating the actions of our limbs and other parts of the body; carrying out such a programmed action would take a lot of time and effort, and the action itself would look clumsy. Our actions and thoughts are subject to verbal programs only in a limited number of situations: in a social environment (school, hospital, theatre), when driving a car on the road, when solving mathematical and logical problems; in most situations, however, our movements and thoughts are regulated not by speech, but by living consciousness.

All and all, experimental confirmation of Vygotsky's thesis that speech thinking is born as a result of the merging of pre-speech thought with pre-intellectual language raises some questions. Specifically, (a) Is pre-intellectual language free from logical structures? (b) When pre-speech thinking turns into speech thinking, does pre-speech thinking as such cease to exist? (c) Is verbal thinking the leading type of thinking in adult's consciousness? and (d) Does verbal thinking solve all the problems facing an adult person in life? From the perspective of the theory of living consciousness, the answer to all these questions is 'no'.

As far as it concerns Question (a), when a newly born baby hears the word 'cat', for him or her this is nothing but a string of non-sensible sounds accompanied by visual patterns, one of which later becomes a special kind of animal and another an adult person who elicits the sounds. Similarly, when a preschool child is shown a letter 'a', he or she doesn't know that this is a letter. What the child is seeing is a visual pattern accompanied by a special sound and an image of an adult producing the sound. It is the innate living consciousness in the form of sensation, memory and associative thinking that allows the baby and the child to put things together and, with some practice, begin to perceive the pattern 'cat' as a name of an animal and the pattern 'a' as a letter included in the context of other patterns ('b', 'c', 'd', etc.) and gradually develop the ability to speak and then read.

When babies create their first words, like 'cat', 'mama', and 'spoon', these words are already under the laws of formal logic, such as identity, non-contradiction, and excluded middle. Specifically, the baby understands that 'mama' is called mama all the time, that this lady is called mama or not mama but not both, and that this lady cannot be called mama and cat simultaneously. Conceptually, language forms as a system of words whose meanings fit to each other according to the laws of logic (see Chapter 1, and Fig. 1.1).

However, the formation of a word meaning itself is a completely different psychological process from the formation of language. Living consciousness, through its property of participation, allows the baby to identify the sound pattern 'cat' with the visual image of a four-legged animal covered with fur. Such identification gives meaning to a meaningless combination of sounds, and thereby turns this sound combination into a word, but at the same time violates the logical laws of identity and contradiction, turning a series of meaningless sounds into a visual image, and vice versa, identifying a visual image with a chain of sounds. From this it is clear that the emerging objectified consciousness in the form of early language does not replace the innate living consciousness. Quite the contrary: innate living consciousness makes the emergence of early language possible.

Understanding this changes the way we think about early speech. If according to L. S. Vygotsky the initial stages in the development of speech are not associated with intelligence, then the origin of language from living consciousness makes it obvious that early language is already subject to the laws of formal logic, and therefore includes logical intelligence. In other words, language and intelligence mediated by logic appear simultaneously as a single whole—as a form of objectified consciousness 'attached' to the body of living consciousness thanks to the magical law of participation—the ability of living consciousness to identify sound patterns with visual ones and thus turn these patterns into words. Somewhat later, according to the same magical law of participation, written speech appears.

To summarize, what L. S. Vygotsky presented as the natural psychological functions is the original form of living consciousness. The very process of 'attaching' the social content in the form of early speech to living consciousness is described by L. S. Vygotsky as the 'law of formation of higher psychological functions'. According to this law, any higher form of behaviour appears twice in its development—first as a collective form of behaviour, as an inter-psychological function, and then as an intra-psychological function. With a certain approximation, inter-psychological forms of behaviour can be associated with external objectified consciousness, and intra-psychological forms with internal objectified consciousness. The very process of transforming 'intra' into 'inter', designated by L. S. Vygotsky as 'interiorization', is nothing more than another manifestation of the magical law of participation—the ability of living consciousness to identify with external stimuli, in this case—with words that came from society, literally turning an external sound pattern into an internal image of this pattern—inner (private) speech.

Participation explains the psychological mechanism of interiorization, and also why learning foreign languages becomes more difficult with age. The point is that if interiorization were a simple translation of the external into the internal according to the principle of 'accumulation', then the effectiveness of interiorization would not depend on the content of what is interiorized. On the contrary, participation assumes that identifying the sounds of a word with an object, and then identifying the 'external word' with its internal equivalent, brings into living consciousness the logic of the external world, which is subject to the laws of identity and contradiction. If a given type of animal is identified with the Russian word 'sobaka' then identifying it with the English word 'dog' means violating the law of contradiction (A or not-A). When children master their native language, this problem does not arise, but at an older age participation increasingly encounters resistance from already-formed logical thinking, which makes participation difficult. As L. S. Vygotsky rightly assumed, in adolescence, memory becomes conditioned by logical thinking (see the next section of this chapter). Such logic-mediated memory promotes better memorization of new ideas but inhibits the memorization of other designations for the same object, in particular, the acquisition of foreign languages.

Answers to the rest of the questions follow from the answer to Question (a). Specifically, the lower mental functions don't disappear because of the formation of the higher ones (b). It is obvious that, along with verbal thinking, an adult also has a non-verbal one, such as artistic thinking, which is carried out not in words, but in images and intuitive insights and is a leading type of thinking in creative activities (c). And it is equally evident that not all problems in life are solved by using verbal thinking, for example, in the field of emotional relationships (d).

To reiterate, verbal thinking has a formal logical structure because words are realized in meanings, and meanings are connected in sensible strings by syntactic and semantic links that are culturally fixed in language. For instance, one can say, 'cats chase mice', but the phrase 'mice chase cats' violates the common-sense logic. In other words, there are no gaps between these 'syntactic/semantic synapses' where something new and creative, such as imagination, can penetrate. Along with verbal thinking, logic begins dominating all higher mental functions, for example, memory [6–8].

It is important to emphasize that with the emergence of speech consciousness, the innate pre-speech forms of consciousness do not disappear, but continue to function and develop alongside their higher mental forms throughout a person's life span. These forms are what in this book is referred to as living consciousness.

2.3 Living Consciousness and Cultural-Historical Approach

As follows from the previous section, living consciousness has properties opposite to properties of higher mental functions. Living consciousness is innate, unmediated by language, involuntary, and not divided into separate functions such as thinking or perception. Depending on circumstances, living consciousness reveals itself in the form of perception (e.g., when seeing a disgusting object spoils the taste of our food), emotion (when we experience real emotions while watching a movie), imagination (e.g., when we see an animal turning into a person in a night dream), and others.

To elaborate on the aforesaid, when we talk about higher mental functions, we distinguish between such mental processes as sensation, perception, imagination, thinking, intentions, feelings, emotions, will, faith and many more. From the perspective presented in this chapter, these distinctions are important rational constructions (RC), applied to living consciousness, or some kind of 'scaffolding', that allows us to meaningfully reflect on the living consciousness by using our objectified consciousness. We have to build the RC in order to put our intuitions about our living consciousness in the flesh of words and concepts, with the aim to reflect at leisure on our living consciousness and communicate about our living consciousness with others (and here the higher mental functions come into the play). However, our intuitive observation of our living consciousness tells us that the RC is an artificial scaffolding that can be demolished, exposing our bare living consciousness—the magical ability of our inner mind to adopt any of the mentioned mental processes depending on current circumstances. When describing the bare living consciousness, we have to speak metaphorically rather than in rational and logical terms. One suitable metaphor is that of a polyhedron, with each facet representing the mental ability designated as imagination, will, perception, and other RC; you can't eliminate any of the facets without changing the whole polyhedron. But this metaphor is still too rationalistically biased. Another metaphor could be that of a multiheaded monster, which displays its heads depending on which of them is needed in current circumstances, while other heads are hidden but ready to be used when appropriate. For example, if you, when eating a tasty meal on the veranda of a respectable restaurant, see in the garden a smelly corpse of a dead animal, you may experience the feeling of disgust, and in this particular manifestation of your living consciousness sensation (smelling the foul odour), perception (seeing the corpse), thinking (finding a proper word for what you are seeing), imagination (imagining that you are eating the corps), affect your intention (to stop eating}, will (rejecting the intention to stop eating) and all other manifestations of your living consciousness, but the 'head of the monster' that you are experiencing at that very moment is the unpleasant sensation of disgust, with other 'heads' laying hidden in the subconscious. It is the living consciousness that helps a disabled person to move a prosthetic hand via brain-computer interface, and not- intention or imagination, although we need these terms in order to rationalistically grasp the magical effect of living consciousness on the physical matter (the brain EEG, see Chapter 5 for more).

The contrast between living consciousness and higher mental functions can be best traced in L. S. Vygotsky's analysis of the imagination [8]. L. S. Vygotsky builds his concept of the imagination on the basis of 'Essay on creative imagination' by Th. Ribot [9], in which the mechanism of the imagination is described as mental actions of dissociation (decomposition) of some realistic images and then subsequent recombination of the resulting elements into novel images by associations based on cognitive or emotional resemblance, contrast, contagion or participation. These combinations are psychological underpinnings of what in anthropology is called the laws of magic, such as sympathy, contagion, and participation [10, 11]. Along with creative imagination based on the laws of magic, Ribot described the psychological mechanism of reconstructive imagination. Reconstructive imagination takes its material from the existing objectified consciousness (e.g., culturally conditioned images, concepts, and theories) and, having reshuffled them, returns some of its products back to reality. Reconstructive imagination is a higher mental function, and in connecting the development of reconstructive imagination with the development of speech L. S. Vygotsky is right. When we want to imagine an object that we have never seen (e.g., a Martian desert), we consciously combine images of objects known to us to create a plausible image. But the combination of known pieces will have to conform to the logic of language. In contrast, creative action must happen before it gets converted into words and concepts. Be it otherwise, the words would push thinking along the line of culturally embedded gutters, which resist violation and thus prevent crucial novelty from emerging. This means that the very action of production of fundamental novelty must be mostly subconscious, internal, and invariant throughout culture and age. In other words, while reconstructive imagination is an advanced form of plagiarism, creative imagination is a form of living consciousness. Examples of the work of creative imaginations can be found in our night dreams, in which, along with regurgitation of ordinary everyday experiences, the impossible and unthinkable combinations of images could happen. Most of these combinations are useless for practical application, but some might give helpful insights. Thus, according to some literary sources, a German chemist August Kekule invented a molecule of benzene by imagining the ouroboros—a snake biting its own tail [12].

Despite living consciousness is innate it develops, both in phylogeny and ontogeny. A monkey jumping in trees when pursued by a predator is a demonstration of living consciousness in its evolutionary early form. Even if the monkey could speak it wouldn't have time to consciously decide which of many branches of the tree ahead are capable to keep its weight and which aren't. The monkey decides instantly, and each decision is vital as a single mistake means falling down and death. This is a model of how creative thinking works in people, when instead of the branches they have images and ideas to 'jump to'.

In ontogeny, the development of living consciousness unfolds in the form of the growing in awareness, volume and the number of manifestations. This contrast with the development of higher mental functions, which occurs in the form of restructuring of connections between the system consisting of these functions. So, with regard to the development of thinking and memory, L. S. Vygotsky writes: "If the thinking of a pre-adolescence child was based on memory and thinking meant remembering, then for a teenager memory is based mainly on thinking: remembering is, first of all, looking for what is needed in a certain logical sequence" [6, p. 120]. This restructuring of connections between memory and thinking, expressed by L. S. Vygotsky theoretically, later was confirmed experimentally [7]. On the contrary, living consciousness is not a system of individual functions, but a unity of interconnected mental processes flowing into each other—the aforementioned 'psychological polyhedron' with such 'facets' as sensation, perception, thinking, imagination, feeling, emotion, will and faith; therefore, the development of living consciousness occurs not in the form of a restructuring of connections, but in the form of an increase in the number of facets, awareness and volume of such a 'polyhedron'. In young children, their living consciousness is mostly unconscious, but later in life much of it becomes conscious subjective experience (see Fig. 1.1). An example of living conscious growing in volume is the growth of the number of things that evoke the feeling of disgust. Initially disgust is elicited by entities that are 'inherently disgusting' [13], but later the number of such entities grows because of cultural conditioning [14]. Another form of living consciousness—imagination—expands from covering the nearest children's environment to later include the whole universe. With age, the number of forms manifesting living consciousness increases as the initial forms of living consciousness diversify: for example, the feeling of attachment to close adults can expand and include attachment to other people, and faith in Santa Claus can develop into faith in the universality of the laws of nature, a single God or monstrous totalitarian media narratives (see Chapter 9). In other words, living consciousness is a basic tree of consciousness, from which higher mental functions branch out, but the basic tree keeps growing throughout the life span (Fig. 2.1).

On a slightly different note, the fundamental role of living consciousness in psychological development is also evident in the structure of social life. Thus, the belief that human relations can be built on rational grounds, which dominated in the nineteenth century, was undermined in the twentieth century (Chapter 6). Today we almost reached the quarter of the twenty-first century but still there are wars and millions of educated rational people fall prey to fraudsters in the domains of information security, economics, and politics (see Part 3 of this book). This suggests that

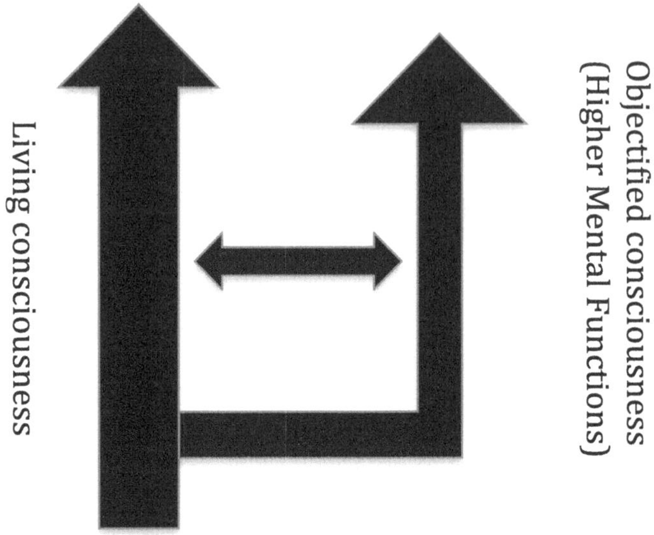

Fig. 2.1 The development of Living consciousness and Objectified consciousness (Higher Mental Functions)

in Vygotsky's lifetime, the role of higher mental functions in human relations was overestimated, while the role of living consciousness was underestimated. Conceptually, naming living consciousness 'lower mental functions' diminishes the role of living consciousness by treating it as inferior to 'higher mental functions' in terms of complexity and importance. In reality, the studies of precocious cognitive abilities of newborns and young infants showed that structurally living consciousness is as complex as higher mental functions [5, 15], but what makes living consciousness special is that after the development of higher mental functions living consciousness retains its pivotal role in human relations (see Fig. 2.1).

To reiterate, the characteristic features of living consciousness are its non-dividedness, the ability to transcend laws of formal logic, and the sense of spontaneity. When we feel, remember, think, wish, or believe, we do not use any specific mental process: on the contrary, all mental processes known to us are involved in every act of living consciousness. Living consciousness is innate, but when higher mental functions are formed, they stay connected with living consciousness through language.

As inner speech, language is a part of internal objectified consciousness: whereas higher mental functions are dominated by language, in living consciousness language is at the service to intuitive revelations and creative imagination.

Most applications of the cultural-historical approach grew out of operating with higher mental functions. For example, in Russia, these applications include the concept of the orientational basis for action by P.Ya. Galperin, school programs for teaching generative concepts by D.B. Elkonin and V.V. Davydov, neuropsychological methods by A. R. Luria, studies of the development of voluntary movements by A.V. Zaporozhets and others. These methods worked, but the properties of living consciousness remained a mystery. The purpose of the cultural-historical approach is to manage living consciousness by directing it along certain prearranged channels, using higher mental functions as 'psychological tools'. But having put all its weight on developing the 'psychological tools', the cultural-historical approach is missing out on studying the very subject these 'tools' are intended to handle—the living consciousness.

2.4 Living Consciousness and the 'Paradox of Learning'

The discoveries of infants' precocious abilities challenged modern theories of cognitive development, including the cultural-historical approach, in one more way. Specifically, Vygotsky's concept of cognitive development (like that of Piaget), by and large, can be exhaustively apprehended within the theory of 'tabula rasa' (the blank slate). According to this theory, which goes back to Aristotle [16] but most clearly was elaborated by John Locke [17], infants are born without built-in knowledge, and their knowledge comes from experience through learning. Both Vygotsky and Piaget acknowledged that newborns had some innate abilities (i.e., reflexes, sensations, perceptions, mnemonic ability, and movements) yet insisted that knowledge comes to the child through learning 'from nature and society'—a unique combination of following the adults' instructions (Vygotsky's emphasis) and independent exploration (Piaget's emphasis). By sharing the 'blank slate' view on cognitive development Vygotsky and Piaget got trapped into what is known as Plato's 'paradox of knowledge' [18]. In Plato's dialogue Meno, Meno argues that in order to be able to learn anything a person has to already know properties or attributes of what he or she is going to learn, otherwise the person wouldn't be able to set up targets for his or her learning, and even if he or she accidentally came across the target, the person wouldn't be able to recognize it. Alternatively, if the person does know the properties and attributes of the target that he or she is aiming to learn, then the person knows the target and there is no point in learning. Either way, there is no point in learning (read 'in cognitive development'). To escape this paradox, Socrates put forward his theory of anamnesis, suggesting that the soul is immortal and repeatedly incarnated; knowledge is present in the soul from eternity, but each time the soul is reborn in someone else's body the knowledge is forgotten in the shock of birth. What appears to be learning is in fact remembering of what one has forgotten. A teacher is an assistant helping the child to recover temporarily 'incapacitated' knowledge, rather than the instructor and model for emulation. Socrates illustrated this theory by asking an uneducated slave boy questions about a geometrical theorem. At first, it appeared that the boy did not know the theorem, but by asking leading questions Socrates was able to help the boy to come to the true answer. This was supposed to demonstrate that the boy was not taught the theorem via instruction but reached the true knowledge by recollecting what he had already known but forgotten.

In the light of modern psychology, Socrates' theory resonates with the concept of L. S. Vygotsky about the 'zone of proximal development'. The zone of proximal development of a child is the distance between the level of his or her actual development, determined by the tasks the child can solve independently, and the level of the child's possible development, determined by the problems solved by the child under the guidance of adults. However, assuming that there is a barrier between the natural psychological functions and the higher psychological functions, L. S. Vygotsky inevitably encounters the problem of a qualitative gap in cognitive development: How do lower psychological functions turn into higher psychological functions if complexity is not inherently present in lower psychological functions? L. S. Vygotsky solves this problem by introducing the descriptive term 'maturation', which he borrowed from biology but the psychological mechanism of which remains unclear.

The discovery of the precocious cognitive abilities of infants removes the problem of the qualitative gap in development, while simultaneously confirming the theory of anamnesis. These findings indicate that newborns and prelinguistic infants have innate cognitive skills comparable in complexity to those of older children and adults, providing continuity in cognitive development. It seems that cognitive development unfolds not as a process of a qualitative leap 'from simple to complex', but as a process of gradual transformation of prelinguistic cognitive abilities into language-mediated forms. This transformation can include growing awareness, voluntary control, linguistic mediation, and linking isolated abilities into a system. In some features, cognitive development as anamnesis fits Vygotsky's model of cognitive development as a transition from lower to higher mental functions. For the concept of development as anamnesis, culture and the educational system retain their significance, but their role in cognitive development must be reconsidered. Culture is no longer the 'keeper' of knowledge that feeds the knowledge into the mind of the child through education. Rather, culture is a master that helps newborn babies to convert their innate cognitive abilities into knowledge.

All and all, the discovery of the striking complexity of lower mental functions challenges the 'tabula rasa' concept of cognitive development and helps Vygotsky's and Piaget's theories of development out of the paradox of knowledge. What must be added is that the living consciousness doesn't come under the command of higher mental functions but continues as a separate line of development along and in parallel with the development of objectified consciousness known as higher mental functions (see Fig. 2.1).

As already mentioned, the socially designed teaching aims at directing living consciousness along the culturally designed gutters. But living consciousness is like water—with the help of mechanics, water can be directed anywhere, forced to beat in fountains, but the structure of water remains unknown. Mechanics isn't enough where chemistry is needed. That is why the study of living consciousness remained predominantly in the hands of writers, artists, and philosophers. However, in psychology too there have been attempts to describe and investigate living consciousness.

2.5 Studies of Living Consciousness in Psychology

The main difficulty in studying living consciousness is finding a method to study something that does not have an objective structure and a stable state. As stated above, in any act of living consciousness sensation, perception, memory, thinking, imagination, desire, will, and faith are involved simultaneously. For example, in his 'Little Book of Great Memory' A. R. Luria described the phenomenon of synaesthesia using the case of the mnemonist Solomon Shereshevsky [19]. In one episode, Shereshevsky reported that, having approached an ice cream kiosk, he asked the saleswoman what kind of ice cream she had. "Fruit ice cream", she said. But she answered in such a tone that Shereshevky felt as if a whole bunch of black coals flew out of her mouth and he could not bring himself to buy the ice cream after she answered that way. This example shows that in living consciousness there are no separate modal forms of perception, but all forms of perception are merged with each other. In everyday life, synaesthesia is evidenced by the trivial phenomenon of disgust: if while dining we remembered the rotten smell of a dead animal we had recently seen, the memory could spoil the taste of food we are eating.

In Soviet psychology, one of the early attempts to study living consciousness by experimental means was A.N. Leontiev's research on the formation of 'skin photosensitivity'—an attempt to show that human perception is essentially non-modal [20]. Like many similar studies of 'anomalous phenomena', this study stands apart from the cultural-historical approach toward the psyche, for example, from the study on the development of higher forms of memorization published in the same book.

A fundamental step in A.N. Leontiev's study was his choice between the 3rd person and the 1st person perspectives of reporting the data. The author explained his choice of the second paradigm by his adherence to the tradition in psychology, which requires "... to use a subjective criterion when establishing the facts of sensitivity" [ibid, p. 70]. However, taken in a broader theoretical context, the author's choice of the second paradigm shows an implicit understanding of the fundamental fact that the explanation of external reality, whether it be a physical object or a mental phenomenon, begins with the living consciousness, in the form of self-observation. To understand this, it is enough to open a school textbook on physics and see that the basic concepts of physics are defined through the student's observation of his or her living consciousness. For example, mass is defined as 'quantitative measure of inertia.... the resistance that a body of matter offers to a change in its speed or position upon the application of a force' [21], but a student can understand this only if he or she had personal experience with the term 'resist', 'change', and 'force', which the student obtained in different life situations. In order to understand the definition of electric charge as 'basic property of matter carried by some elementary particles that governs how the particles are affected by an electric or magnetic field' [22]) the student must learn from his or her personal experience such notions as 'matter', 'property', and 'affected'. This means that living consciousness is implicitly included in the concepts of physics and cannot be derived from or reduced to these concepts without falling into a logical circle. Likewise, in psychology, the phenomenon of sensitivity as a form of living consciousness cannot be extracted from objective observations only but must include subjective reports, and only on that basis it can be subsequently projected onto other people and animals.

This is exactly what A.N. Leontiev does. To establish experimentally that the sensation of light can arise on the basis of an organ that is not adapted by evolution to respond to the impact of light waves, along with registering the participants' behavioural reactions A.N.Leontiev uses the participant's verbal reports. Notably, the verbal reports were crucial evidence of the fact that the participants' behavioural responses were not mere reflexes but exhibited their living consciousness in the form of sensations [23].

2.6 Living Consciousness and Cognitive Psychology of the Mainstream

Cognitive psychology imitates physical sciences. When psychologists report 'normal' psychological effects, the results are assumed to be stable over the duration of the experiment and over time after the experiment. This means that the subject matter of cognitive psychology is the innate objectified consciousness, in the form of cognitive skills in the domains of thinking, memory and perception (see Fig. 1.1).

In contrast to objectified consciousness, where the subject and object are separated from each other, in living consciousness this separation is absent and living consciousness of different people can communicate with each other [24], influence processes in the external world [25, 26] or 'see the nearest future' [27, 28]. Although phenomena of living consciousness such as remote viewing or psychokinesis are being actively researched, they stand apart from mainstream psychology because they have a low degree of reproducibility. The effects of living consciousness can decrease significantly with testing time even if the experimental conditions remain constant, and a person can manifest these effects and then lose this ability.

In addition to providing evidence for the 'anomalous phenomena', the tasks of studying living consciousness include the investigation of its functioning: the study of subjective space, time, and causality. In this direction, certain results have been achieved in a relatively new area of research - magical thinking [29, 30] (see Chapter 3 for more on that). Finally, one more form of studying living consciousness is immersion in special states of consciousness with the help of psychotropic substances [31].

2.7 The Mixed Blessing of Psychotherapies

As for the practical impact on living consciousness, it has existed since the dawn of times, in the form of magic and religion, and in medicine—in the form of the placebo effect. Today, the influence on living consciousness is known under the names of psychotherapy, psychological help, and suggestion. A 'therapist' can be a religious leader, a magician, a politician, a doctor, a psychotherapist, and anyone who tries to influence others living consciousness with words, gestures, images, or otherwise. Any psychological therapy works on the basis of 'participation'—the client's involuntary identification with the therapist's message (see Chapters 7, 8, and 9 for the elaboration on that). The property of living consciousness to experience participation means that along with positive aspects, the therapy harbours serious dangers. Having excluded from the category of therapist outright scammers, let us outline some of the side effects that lie in wait for therapists with positive intentions for their clients.

Metamorphosis. While working with a client, the therapist may be tempted, under the guise of psychological help, to solve his or her own personal problems, both financial and psychological (e.g., to experience the sense of his or her power and control over the life and behaviour of a client). In this case, the therapy from psychological assistance turns into a magical manipulation of consciousness (see Chapter 7 for more detail). Since the 'red line', which the therapist crosses at such metamorphosis, may be imperceptible both for the client and for the therapist, it is necessary to develop forms of control and self-control that prevent the therapist from crossing this line.

Suspension. Another danger that awaits the therapist working with living consciousness is the absence of external supports for action: while the therapist works with the client's magical thinking, the therapist's own thinking is subject to the laws of magic as well. For example, a physicist or a doctor has the laws of physics or biology for orientation in dealing with matter or the human body. A psychologist working within the framework of a cultural-historical approach has the same external supports in the laws of logic, scientific concepts and social norms of behaviour. The therapist of living consciousness has no such external supports: since living consciousness doesn't conform to objective laws the therapist has to rely on his or her own living consciousness with which the client may or may not identify. In other words, in working with living consciousness much depends on the therapist's intuition, and this makes the result of the therapeutic process only probabilistically predictable.

Overdose of influence. According to psychoanalysis, any conscious action of a person requires the expenditure of psychic energy [32, 33]. Psychic energy should not be confused with physical energy, which we expend when performing physical exercises. For instance, after a difficult exam we may feel tired like after hard work. Studies have shown that psychic energy is not a metaphor but a real force that can be measured by cardiovascular and biochemical parameters such as heart rate or blood glucose levels [34, 35]. Experiments revealed that any act of self-control causes blood glucose levels to drop below optimal levels, making subsequent attempts at self-control difficult. The expenditure of psychic energy from the person's limited reservoir of such energy occurs in any activity that includes an effort to self-control, such as the process of coping with unpleasant thoughts about death. In a series of experiments, participants were asked either to write a story about death or to write a story on a neutral topic. It turned out that those who had written about death performed significantly worse on subsequent self-control tasks than those who had written on a neutral topic [36, 37]. Since the therapy of living consciousness requires a large expenditure of the patient's psychic energy, excessive intervention into the patient's living consciousness can lead to depletion of psychic energy and, as a result, to depression or, in patients with neurological disorders, to worsening of their conditions.

Energy vampirism. The psychic energy of living consciousness can be transferred to others through submission, praise, and other forms of psychological interaction. Thus, a person who has received an award or recognition from people experiences a surge of psychic energy, and vice versa, humiliation or ignoring the person's achievements could get the person into a state of depression. Since the therapist needs the unconditional trust of the client, there is a danger of exploiting this trust. The client's sense of trust in the therapist can develop into boosting the therapists' sense of his or her own power over the client's feelings. Having gone out of control, this sense of power could turn the therapist from a person giving psychological help into a person using the client's adoration for replenishing the therapists energy reservoir—the phenomenon known as 'energy vampirism' [38].

Privacy violation. A defining feature of living consciousness is its privacy. No one, unless he or she has a unique ability of telepathy, has access to someone else's living consciousness. Therefore, access to one's living consciousness, even if it is given voluntarily by the client, can turn out both good and bad. The person may initially feel relieved by sharing their secrets with the therapist but will subsequently feel frustrated.

Altogether, practical work with living consciousness without the knowledge of living consciousness' properties is a 'game without rules', or a kind of art. But in contrast to art, which is the objectification of the artist's living consciousness, psychological therapy can irreversibly affect the fate of a person.

2.8 Personal Experience

In the 1970s, I was a participant in the first psychological training in the USSR (the so-called T-group), which was conducted by a Polish psychotherapist. I was amazed by how easy it was for a therapist to cross the line between helping and manipulation. It was evident that not all members of this group felt comfortable, and some went out with depression. Subsequently, I repeatedly conducted demonstrations of the work of T-groups in a psychological workshop with students, highlighting both gains and losses. In 1980s I attended lectures by Carl Rogers, who came to Moscow to teach his therapeutic method, after which I studied the 'client-centred' psychotherapy in more detail. At about the same time I observed the effects that television sessions by the psychotherapist Anatoly Kashpirovsky had on people. Some viewers reported improvement through a placebo effect, others felt nothing, and my mother, who had previously had a stroke, experienced a sharp deterioration in her condition. I watched the TV shows by the faith healer Alan Chumak, aimed at eliciting the placebo effect. These personal experiences made me aware of the dangers that psychological help harbours, and the need for a practical psychologist to be familiar with properties of living consciousness.

Finally, I conducted experimental research, alone and in collaboration, on the impact of living consciousness on external reality [26, 27] and on the impact of external reality on living consciousness [39–43]. These experiences made me re-evaluate the role of the studies on 'psi' extraordinary phenomena and the limitations of the cultural-historical approach towards living consciousness.

2.9 Limits of Applicability

The cultural-historical method aims at managing living consciousness with the help of objectified structures of consciousness—symbolic formations such as speech, logical thinking, scientific concepts, social norms, or culturally developed perceptual patterns. These structures are used as 'psychological tools' that shape the work of living consciousness. This shaping leads to controversial results: while adapting to the requirements of the social environment necessary for communication with others, the living consciousness loses its freedom, spontaneity, and ability to generate creative innovations.

Within the framework of the cultural-historical approach, it is assumed that a person's living consciousness is determined by culture: language, formal logic, science education, and social norms. The cultural-historical method works with internal objectified consciousness via constructing the external 'scaffolding' in the form of methodical guides and algorithms. This method is effective at teaching scientific knowledge at school, developing cognitive skills, restoring lost cognitive functions in patients, and shaping socio-moral forms of behaviour based on external control. However, this method is inapplicable to studying living consciousness per se: subjective experiences, motivation, emotions and creative thinking.

In contrast, the method applicable to the study of living consciousness can be called the method of releasing influence—an experiment based on a subject's self-observation over his or her subjective experiences and 'objectivising' these experiences in conditions where the subject can freely structure reality. Using an analogue from cosmology, we can say that this method creates conditions when living consciousness is considered as a 'black hole' emitting 'Hawking radiation' in the form of creative impulses. The method of releasing influence considers a person as a source of free and spontaneous activity—as a subject who is literally 'making himself or herself'. This method originated in psychoanalysis as the method of free associations and analysis of fantasies and dreams and later developed into other projective techniques.

Unlike the cultural-historical method, which aims at influencing living consciousness through coercion, the method of 'releasing influence' allows the living consciousness to unfold in the form of a game, fantasizing, or free observation. For example, studies have shown that watching fragments from Harry Potter films that contained elements of magic facilitated creative thinking in children [39], watching magical effects in commercial clips improved recognition of commercial brands [40], and the game of discriminating between real and fantastic images helped to improve the fantasy/reality distinction [41]. In educational psychology, the way of releasing creative potentials of living consciousness is through the altruistic communicative style, which, if applied consistently in a classroom throughout a considerable time, has a boosting effect on children's non-pragmatic moral behaviour, critical thinking, and artistic creativity [42, 43]. This does not mean that the impact of culture on living consciousness is unimportant, but such an impact influences not living consciousness as such but the degree of its influence on human actions [44].

The limits of the cultural-historical approach to living consciousness were acknowledged by some of the advocates of this approach. For example, in his latest book 'Activity. Consciousness. Personality', A.N. Leontiev emphasizes that the culturological approach is based on the recognition of direct determination of consciousness by cultural factors—the so-called 'postulate of immediacy'. "There is a simple substitution: the place of the world of objects is now occupied by the world of signs and meanings developed by society. Thus, we are again faced with the two-term scheme S- > R, but only the stimulus is interpreted in it as a 'cultural stimulus'" [45, pp. 35–36]. In contrast to this approach, A.N. Leontiev claims that "Consciousness as a form of mental reflection... cannot be reduced to the functioning of meanings learned from outside, which, unfolding, govern the external and internal activity of the subject. The values and the operations wrapped in them taken in their abstraction from the internal relations of the system of activity and consciousness, are not at all the subject of psychology. They become this subject only when they are taken in these relations, in the movement of their system" (ibid., p. 69). One can assume that the "internal relations of activity and consciousness" in the terms of A.N. Leontiev is the equivalent of living consciousness. However, A.N. Leontiev is not always consistent in describing living consciousness. In the same work, he writes, "Psychology has long described and studied the dependence of perception, representation, thinking ...on needs, motives, attitudes, and emotions. At the same time, it is very important to emphasize that such partiality is itself objectively determined ..." (ibid., p. 24). However, if living consciousness is objectively determined, then it is not living consciousness by definition. Rather, what A.N. Leontiev wanted to express here is that not all in our living consciousness can be consciously controlled. Many perceptions, thoughts, and feelings in our living consciousness are not responsive to direct orders of our internal Self. The objective factors of our life, such as cultural affiliation, social status, profession, age, education, and relationships with other people affect not our living consciousness but our external Self in the form of character, temperament, attitudes, habits and other 'fossils of subjectivity'. As for our internal Self, it constantly revisits our existing 'fossils', by confirming, changing, or rejecting them and building our new external Self.

Nevertheless, A.N. Leontiev claims that "The special function of sensory images of consciousness is that they give reality to the conscious picture of the world that opens up to the subject" and that "it is precisely thanks to the sensory content of consciousness that the world appears to the subject as existing not in consciousness, but outside his or her consciousness – as an objective 'field' and the object of the subject's activity" [ibid, p.64]. In essence this statement means that any impact, both physical and cultural, exists only if it is a part of the person's living consciousness. If a person is not conscious of the impact, this impact does not exist in his or her private universe. If a person dies, then the outside world with all its majestic grandeur of galaxies, evolution, cultures, and history dies as well. Of course, we want to believe that the world remains, but this is not knowledge but a belief that can't be empirically validated.

All and all, understanding external reality as the objectified consciousness resulting from the fusion of living consciousness and reality independent of a person puts a fundamental limit to the cultural-historical approach: Being an effective method of studying internal objectified consciousness, the cultural-historical method stops at the 'magical gate' guarding the entrance to the realm of living consciousness (see Fig. 1.1).

In the next chapter, we will consider magical thinking and magical experience as dimensions of living consciousness.

References

1 https://www.vocabulary.com/dictionary/psyche
2 Vygotsky, L. S. (1934). Mishlenie I Retch (Russ). https://fip.kpmo.ru/es_ru/0_publication_109_1.pdf
3 Subbotsky, E. V. (1976). Psychology of partnership relations in preschoolers (In Russ). Lomonosov State University Publ.
4 Luria, A. R., & Subbotsky. E. V. (1978). Zur fruhen Ontogeneze der steuerden Funktion der Sprache. Die Psychologie des 20 Jahrhunderts (pp. 1032–1048). KinderVerlag.
5 Subbotsky, E. (1996). Vygotsky's Distinction between lower and higher mental functions and recent studies on infant cognitive development. *Journal of Russian and East European Psychology, 34*(2), 61–66.
6 Vygotsky, L. S. (1930). O psykhologicheskih systemakh (Russ). Collected works (Vol. 1). Pedagogica.
7 Subbotsky, E., Chesnokova, O., & Greenfield, S. (2002). Inference-based memory failures in children and adults. Poster presented at the 3rd Tsukuba International Conference on Memory, University of Tsukuba, Japan. https://www.lancaster.ac.uk/staff/subbotsk/Tsukuba%20pres,%202002.pdf
8 Vygotsky L. S. (1932). Voobrazheniye I tvorchestvo v detskom vosraste (Russ). http://школаискусств1.екатеринбург.рф/file/1e3f72bec064cf65aa12e22e6be4a8ae
9 Ribot, Th. (1906). Essay on Creative Imagination. https://www.gutenberg.org/cache/epub/26430/pg26430-images.html
10 Lévy-Bruhl, L. (1923). *The primitive mentality* (English translation). Alcan.
11 Lévy-Brühl, L. (1985/1926). How natives think. Princeton University Press.
12 https://dodona.ugent.be/en/activities/1633022739/
13 Curtis, V., de Barra, M., & Aunger, R. (2011). Disgust as an adaptive system for disease avoidance behaviour. *Philosophical Transactions of the Royal Society B: Biological Sciences, 366*(1563), 389–401.
14 Haidt, J., Rozin, P., McCauley, C., & Imada, S. (1997). Body, psyche, and culture: The relationship between disgust and morality. *Psychology & Developing Societies, 9*, 107–131. https://doi.org/10.1177/097133369700900105.S2CID144762306
15 Bremner, G. (1994). *Infancy*. 2nd edition. Wiley-Blackwell.
16 Aristotle. (1936). On the soul (De anima), trans. W. S. Hett. In Loeb classical library. Aristotle (Vol. VIII, pp. 1–203). William Heinemann.
17 Locke, J. (1996). *An essay concerning human understanding*, Winkler K. P. (Ed.) (pp. 33–36). Hackett Publishing Company.
18 Day, J. M. (1994). *Plato's Meno in focus*. Routledge.
19 Luria, A. R. (1994). *Small book about great memory*. Eidos.
20 Leontiev, A. N. (1981). Problemy razvitiya psikhiki (Russ). Moscow State University.

21 https://www.britannica.com/science/mass-physics
22 https://www.britannica.com/science/electric-charge
23 Leontiev, A. N. (1947). Ocherk Razvitiya Psyckhiki (Russ). Military Pedagogical Institute of the Soviet Army.
24 Bem, D. J., & Honorton, C. (1994). Does Psi exist? Replicable evidence for an anomalous process of information transfer. *Psychological Bulletin, 115*, 4–18.
25 Dunne, B. J., Nelson, R. D., & Jahn, R. G. (1988). Operator-Related Anomalies in a Random Mechanical Cascade. *Journal of Scientific Exploration, 2*, 155–179.
26 Subbotsky, E. (2013). Sensing the future: Reversed causality or a non-standard observer effect? *SAGE open*, 1–17. https://openpsychologyjournal.com/VOLUME/6/PAGE/81/ABSTRACT
27 Subbotsky, E. & Ryan, A. (2015). Motivation and belief in the paranormal in a remote viewing task. *The Open Behavioural Science Journal, 15*. https://benthamopen.com/ABSTRACT/TOBSJ-8-1
28 Roe, C. R., Cooper, C. E., Hickinbotham, L., Hodrien, A., Kirkwood, L., & Martin, H. (2020). Performance at a precognitive remote viewing task, with and without Ganzfeld stimulation: Three experiments. *Journal of Parapsychology, 84*(1), 38–65.
29 Nemeroff, C., & Rozin, P. (2000). The making of the magical mind: The nature and function of sympathetic magical thinking. In K. S. Rosengren, C. N. Johnson, & P. L. Harris (Eds.), *Imagining the impossible: Magical, scientific and religious thinking in children* (pp. 1–34). Cambridge University Press.
30 Subbotsky, E. (2010). *Magic and the mind. Mechanisms, functions and development of magical thinking and behaviour*. Oxford University Press.
31 Pollan, M. (2018). *How to change your mind: The new science of psychedelics*. Penguin.
32 Freud, S. (1935). *A general introduction to psychoanalysis*. Liveright.
33 Jung, C. G. (1960). *On psychic energy*. Princeton University Press.
34 Benton, D., Parker, P. Y., & Donohoe, R. T. (1996). The supply of glucose to the brain and cognitive functioning. *Journal of Biosocial Science, 28*, 463–479.
35 Fairclough, S. H., & Houston, K. (2004). A metabolic measure of mental effort. *Biological Psychology, 66*, 177–190.
36 Gailliot, M. T., Baumeister, R. F., & Schmeichel, B. J. (2006). Self-regulatory processes defend against the threat of death: Effects of self-control depletion and trait self-control on thoughts and fears of dying. *Journal of Personality and Social Psychology, 91*, 49–62.
37 Gailliot, M. T., Baumeister, R. F., DeWall, C. N., Maner, J. K., Plant, E. A., Tice, D. M., Brewer, L. E., & Schmeichel, B. J. (2007). Self-control relies

on glucose as a limited energy source: Willpower is more than a metaphor. *Journal of Personality and Social Psychology, 92*, 325–336.
38 Csoff, R. (2022). Energy Vampirism—An Issue for Psycho-Social Self-Care and prevention. Academia Letters, Article 4590. https://doi.org/10.20935/AL4590
39 Subbotsky, E., Hysted, C., & Jones, N. (2009). Watching films with magical content facilitates creativity in children. *Perceptual and Motor Skills, 111*, 261–277.
40 Subbotsky, E., & Matthews, J. (2011). Magical thinking and memory: Distinctiveness effect in for TV advertisements with magical content. *Psychological Reports, 109*, 1–11.
41 Subbotsky, E., & Slater, E. (2011). Children's discrimination of fantastic vs realistic visual displays after watching a film with magical content. *Psychological Reports, 112*, 603–609.
42 Subbotskii, E. V. (1983). Shaping of moral action in children. *Soviet Psychology, 22*(1), 56–71. https://www.researchgate.net/publication/270238514
43 Subbotskii, E. V. (1987). Communicative style and the genesis of personality in preschoolers. *Soviet Psychology, 25*(4), 38–58. https://www.researchgate.net/publication/236984323
44 Subbotsky, E., & Quinteros, G. B. (2002). Do cultural factors affect causal beliefs? Rational and magical thinking in Britain and Mexico. *British Journal of Psychology, 93*, 519–543. https://www.researchgate.net/publication/10957312
45 Leontiev, A. N. (1975). Activity, Consciousness, Personality. Politizdat.

CHAPTER 3

Magical Thinking and Magical Experience as Dimensions of Living Consciousness

3.1 Two Dimensions of Living Consciousness

Living consciousness follows not the laws of logic but the laws of magic. Accordingly, two different dimensions of living consciousness can be distinguished—magical thinking and magical experience.

The fundamental feature of magical thinking distinguishing it from logical thinking is that laws of magical thinking transcend the primary laws of classical logic, such as the law of identity [A is A], contradiction [A is not not-A], excluded middle [X is either A or not-A] and sufficient reason [If A then B (A implies B)] [1]. More specifically, in the domain of perception the magical law of sympathy makes one treat physically separate but resembling each other objects (e.g., a person and the person's photo) as identical; similarly, according to the magical law of contagion, two different objects that used to be in physical contact (e.g., a person and the person's cloths) are treated as the same. In the domain of communication, according to the magical law of participation one can take a suggested message for truth contrary to the understanding that this message is wrong (see Chapter 7 for the elaboration on that). In the domain of imagination, the magical law of transformation makes people consciously create impossible combinations of images (e.g., animals speaking human languages or fantastical creatures). For example, in ancient religious cults gods were depicted as creatures in which human and animal features

E. Subbotsky, *The Magic of Living Consciousness*,
https://doi.org/10.1007/978-3-031-54708-9_3

were combined [3], and today some trends in modern art (e.g., surrealism) make perceptual magical thinking their main method of creating alternative visual reality.

In contrast to magical thinking, which unfolds through cognitive processes, in magical experience the same lows of magic work through emotions. In the domain of perception magical experience manifests itself in the feeling of disgust, for example when participants in a psychological experiment rate a piece of chocolate shaped as a disgusting object as less desirable to taste than a piece of chocolate looking like a disc or a muffin [2] Another example of magical experience is the phenomenon of synaesthesia (see the examples in Chapter 2). In the domain of imagination, people have magical experience when they emotionally take fake events for real ones and identify themselves with fictional characters. The defining feature of magical experience is that in the magical experience emotions contradict rational knowledge and logical reasoning. The opposite of the magical experience is typical emotions that are elicited by internal (e.g., the feeling of joy when achieving consciously set goals) or external (e.g., the feeling of pleasure from enjoying favourite food) stimulation. Socially elicited emotions, such as emotions triggered by rewards and punishments, don't' belong to magical experiences as well.

To summarize, within living consciousness logical laws are transcended by magical thinking, which is predominantly a conscious cognitive activity, and by magical experience, which is predominantly a subconscious emotional activity. In contrast, within the objectified consciousness, the cognitive domain conforms to the laws of logical thinking and manipulation with regular objects and events, and the emotional domain comprises typical emotions that accompany rational reasoning, emotions elicited by external stimulation, and emotions simulated but not really experienced (see Table 3.1).

3.2 The Laws of Living Consciousness and Physical Reality

As stated above, in living consciousness different entities or objects are connected not by the laws of logic, but by the laws of sympathy, contagion, and participation [4–8]. One more law of magic is mind over matter, when our living consciousness directly affects matter. An example we find in the Book of Genesis: "And God said, Let there be light, and there was light" (Genesis, 1:3). Magical laws distinguish phenomena of

Table 3.1 Types of Consciousness as a Function of *Dimensions* and *Psychological Domains*

	Types of consciousness			
	Living Consciousness		*Objectified consciousness*	
Dimensions Psychological domains	*Magical thinking*	*Magical experience*	*Cognitive*	*Emotional*
Perception and communication	Identifying separate objects by the laws of sympathy and contagion. Taking a suggested message for truth contrary to the understanding that the message is wrong.	The feeling of disgust. The phenomenon of synaesthesia.	Creation of regular shapes in realistic art. Logical thinking.	Emotions that accompany achievements. Emotions elicited by social reinforcements.
Imagination	Creating non-existing entities via merging parts of different entities in art and religion. Believing that a magic spell could affect events in real life.	Emotional identification with fictional characters and events. Falling in love with a person or an object.	Imagining regular everyday objects (e.g., your car left in the garage for repair).	Simulating emotions without really experiencing them (politeness, diplomacy).

living consciousness from phenomena of objectified consciousness, which is commonly known as physical reality.

In physical reality an object maintains its identity until it is destroyed by another object, and even then it doesn't disappear completely and conforms to the law of conservation of energy and matter. In addition, physical objects are local: whatever happens to one object does not affect another unless through known physical forces. For example, events in a faraway galaxy do not affect events in our galaxy until the light from these events reaches our galaxy, or burning a picture of the person doesn't affect the real person.

On the contrary, in living consciousness phenomena are fundamentally non-permanent and non-local. For example, while strolling in the street we suddenly catch the odour of cherry tree blossom, which reminds us the episode from our childhood when we were stealing cherries from the neighbour's garden, which brings about the memory that the neighbour was a retired sailor, and this thought brings about the dream of sailing on a yacht to a tropical island, etc., and the chain of associations could go on in the most unpredictable and bizarre way. In other words, in our living consciousness thoughts and feelings are magically connected, either subconsciously or consciously, and we don't find this unusual. However, magical thinking becomes more challenging when it turns into magical beliefs.

3.3 Magical Thinking and Magical Beliefs

Whereas magical thinking unfolds inside living consciousness, the belief in magic affects our behaviour. Magical behaviour occurs when the laws of magic, which govern our living consciousness, are 'smuggled' into external reality bypassing the buffer zone of internal objectified consciousness via the route of direct intrusion (see Fig. 1.1). Having done this intrusion into the external world, magical laws can either settle down in the domain of art or proceed into physical reality through affecting our verbal or actual behaviour. If we admit, consciously or subconsciously, that magic can govern the world of physical reality, we turn from magical thinkers into believers in magic, and our rational behaviour turns into magical behaviour.

In the ancient time magical behaviour legally coexisted with rational behaviour. Ancient Romans used to write magical incantations on the plates of led and didn't start important events without good omens. Today science convinced most of us that belief in magic contradicts everyday experience and the fundamental laws of nature. So, it appears that in the modern industrial world magical behaviour is banished from the physical world and is securely locked within our living consciousness, leaving behind the insignificant trace in the form of superstitions and children's games. Nevertheless, psychological research showed that rational people today still harbour the belief in magic.

3.4 Do Children and Adults Believe in Magic?

Psychologists have long noticed that 4- to 7-year-old children often reason as if they believed in magic [9, 10]. What remained unclear though was whether the children could behave in accordance with such beliefs.

To examine that, Moscow children of 4, 5, and 6 years old were individually told a story about a girl who was presented with a magic box [11]. The box could turn pictures of objects into real objects shown in the pictures. When children were asked if such a box could exist in real life, everyone answered 'no'. Next, the children were presented with a similarly looking real box and cards with attractive objects drawn on them. When the experimenter left the room 'to make a phone call', 90% of the children tried to magically turn the pictures into objects and were disappointed when this didn't happen. In other studies, British 4–6-year-old children behaved as if the creature they had imagined appearing in an empty box was really in the box [12], and 3–4-year-old American children were made believe in a new magical character invented by the experimenters [13]. These studies showed that at the age of 4 years most children deny that they believe in magic, but in reality 4 to 6-year-olds behave as if magic had causal powers. Rational adults exhibit magical behaviour as well. When religious people engage in a petitionary prayer or follow the Eucharist ritual in catholic Mass, they act as believers in the supernatural. The question arises whether magical behaviour in modern adults is an institutionally imposed religious ritual, or it can be elicited beyond the religious context?

With the aim to answer this question, modern university graduates and undergraduates in the United Kingdom were asked to imagine that a witch wanted to put a magic spell on them. The witch said that the spell would only work if the person allowed her to put the spell. In one scenario the spell was good and aimed at making the person's future life nice and wealthy. In another scenario, the spell's aim was to make the person a servant to evil forces. In the interview, the participants denied that the magic spell could have any effect on their future lives, but when asked whether they would allow the witch to put the spell about half of the participants allowed the good spell but almost all prohibited the bad one and justified their prohibition by admitting that the bad spell might work [14]. This study suggested that the belief in magic, that we consciously deny, can be activated outside the religious context.

But if magical thinking and the belief in magic are inseparable components of our living consciousness, could these components be used for practical purposes?

3.5 Curiosity Towards Magic and the Motivation for Learning

It is known that learning in humans and animals can be stimulated by extrinsic and intrinsic motivation [15]. Extrinsic motivation is based on basic motives and applied when there is no self-interest in learning. For example, the child can be asked to complete the boring homework first and then allowed to play the computer games. In contrast, intrinsic motivation is based on the student's own interest in the material. Studies have shown that extrinsic motivation negatively correlates with academic performance, whereas intrinsic motivation positively correlates with school grades and performance on special tests at all levels of education [16].

The main kinds of intrinsic motivation for learning are curiosity and desire for exploration. Studies of the exploratory behaviour of animals [17], children [18], and adults [19] showed that curiosity and desire for exploration can motivate learning to the same extent as the desire to receive rewards or avoid punishments. The main motivating factor in exploratory activity is the novelty of the stimulus [20, 21]. However, the novelty as such is not sufficient to motivate exploratory behaviour; for a stimulus to acquire motivating power, it must be not only new but also interesting and attractive. Thus, new foods do not make people and animals want to try them if these foods look unattractive or dangerous to consume [22, 23].

It has long been known that different forms of magic elicit curiosity in people [24, 25]. The difference between a magical event (e.g., the ability to move inanimate objects by just thinking about it) and a novel physical event (e.g., the ability to move objects through a previously unknown physical force) is that a magical event is not only new but also unusual and attractive. Psychological experiments revealed that both children and adults exhibited a much stronger interest in exploring magical events than in exploring equally novel and interesting physical events [26]. This suggests that the interest in exploring magical events can serve as a source of motivation for learning, comparable in strength, or even stronger than the interest in exploring novel physical events.

The question arises, what powers the interest of children and adults to explore magical events? Certainly, when learning is motivated extrinsically, the energy for learning is supplied by the desire to satisfy innate needs, such as the need for food, communication, or safety. In contrast, when the learning material is intrinsically attractive, the energy for learning such material is provided by the desire to explore the material. In other words, the intrinsically attractive material is a self-supplier of psychic energy. Magical events are intrinsically attractive and thus generate psychic energy that powers their exploration. It is possible to assume that exposure to magical events generates the amount of psychic energy which exceeds the amount of such energy needed for exploration of the magical events per se. If this is the case, then the extra amount of psychic energy generated by exposure to magical events can be used for powering other cognitive activities, such as creative thinking and memory.

3.6 Magical Experience as a Generator of Psychic Energy

Psychological studies indicated that any action of living consciousness that involves stress and self-control requires the expenditure of psychic energy from a limited reservoir and reduces blood glucose levels [27, 28]. This energy can be replenished by external sources and subsequently used for facilitating cognitive functioning. Thus, viewing emotionally exciting images increases blood glucose levels and improves memory to a much greater extent than viewing similarly complex and unusual images that do not emotionally affect a person [29]. By analogy, it was assumed that magical experience of watching magical effects in a feature film produces an extra amount of psychic energy that can be used for powering creativity, memory, and learning (see Fig. 3.1).

To examine this, in a series of psychological experiments children were engaged in watching video clips with exciting magical events and then tested on cognitive skills. Research has shown that 4- to 6-year-old children, after having been exposed to clips with magical effects, performed significantly better on creative associative thinking tasks than children exposed to clips from the same movie that didn't contain magical effects [30]. In another study, exposure to films with magical effects significantly improved children's ability to learn the distinction between fantastical and realistic visual displays [31]. Further, the inclusion of commercial

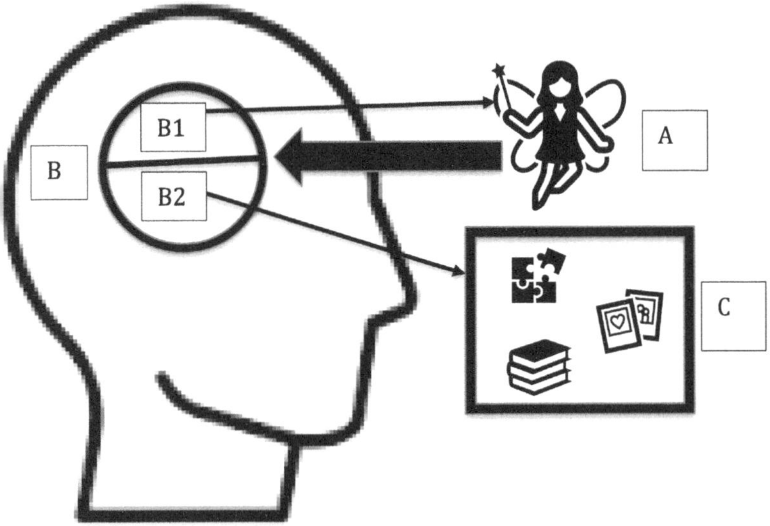

Fig. 3.1 Magical experience **a** generates psychic energy **b**, part of which **b1** self-powers the magical experience, while the extra amount **b2** powers creative thinking, learning and memory **c**

brands in the context of magical events enhanced the ability of adolescents and adults to recognize these brands to a much greater extent than the inclusion of the same brands in the context of equally interesting but non-magical events [32]. This explains the fact that parents and teachers use children's interest towards magic to enhance motivation for learning, by dressing the learning material in the form of fairy tales or magical games. In line with this, some psychologists argue that teaching children magic tricks increases their ability to self-regulation and critical thinking [33–36].

Finally, children's knowledge of the ability of magical characters and events to transcend the boundaries of the known laws of nature and psychology can be used for all sorts of educational games. For example, a typical technique used to teach language to children in primary schools in the United Kingdom is the open-sentence exercise: the teacher covers part of a sentence or word with opaque paper and asks the children to 'read' the closed part, offering the following imaginary situation: "Imagine that you are magicians and can see through paper". The use of such instructions presupposes the ability of children to understand that wizards can do things that ordinary people cannot. In principle, it is possible to create alternative handbooks on scientific subjects, in which familiar things would follow not the laws of nature but the laws of magic instead. Such 'real versus magical' explanations of the same phenomena can be done in the form of art or in the form of 'compare and contrast'. The example of the first form is Jean Effel's «The Creation of the World» cartoon caricature book [37], and the example of the second form is comparing the Big Bang theory of the emergence of the universe [38] to the biblical story of creation. Discussing with children common features and differences between scientific and biblical versions of the origins of the universe or the origins of species might make science lessons more interesting and memorable.

One might say that such 'alternative handbooks' already exist, if not in real life then in fiction books and movies. For example, the «Harry Potter» series can be used for helping to increase motivation in teaching science. When watching the episode with Harry wearing the cloak of invisibility it might be useful to discuss whether the hero himself could see anything at all; indeed, in order to be invisible a person has to be transparent and let all the light through his or her body, but if the person's eyes do not absorb light, the person can't extract from the light any information and therefore is unable to see anything. When the characters of the movie ride broomsticks, one might ask children how it is possible to maintain equilibrium at turning points while sitting on a narrow rod and without having any fulcrum; it might also be useful to discuss issues such as gravitation, inertia, and centrifugal forces. When watching the episode of Harry and friends going through the stone wall to get into the magic land, the children could be asked how the stone wall could let people through without letting light through simultaneously. Discussions like that could help children reflect upon the known laws of physics, biology, and psychology better than direct instructions in those laws.

All and all, magical thinking and magical experience, far from being just a temporary imperfection of children's ideas about the world, have a hidden and not fully appreciated the potential for learning and cognitive development.

3.7 Magical Experience and the Belief in Magic as the Source of Influence and Mind Control

As argued in the previous section, magical experiences can generate psychic energy that can be used to fuel cognitive functioning. However, although magical experiences produce real emotions, they nevertheless are elicited by situational stimuli, such as observation of magical effects on stage or in movies. A much stronger generator of psychic energy could be magical beliefs. If a person not only enjoys observing virtual magical characters but believes that God or a political leader has magical causal powers, the person's belief could generate psychic energy sufficient not only for powering cognitive functions but also for powering his or her real actions. Sadly, the most salient illustration of the capacity of explicit magical beliefs to produce psychic energy comes from the actions of religious extremism.

Despite a decline of religious beliefs in modern industrial societies, many people still consciously believe in God, and religion has a common historical root with magical beliefs [39]. If the energy generated by the people's explicit belief in supernatural powers of their leaders is taken under control by these leaders, it can be used for manipulation with the people's living consciousness. Examples of such manipulation abound in history. Thus, the suicidal heroism that Japanese kamikaze who sacrificed their lives for their emperor was powered by their belief in the emperor's divine origins [40]. Cases of mass suicides of controversial cult members, undertaken on their leaders' orders, are cited in general psychology texts [41, 42]. Anthropological research on suicidal terrorism today suggests that 'sacred values' are at the core of modern political terrorism and that these values supersede economic and other material considerations [43]. Thus, most Palestinian suicide bombers do not differ from the average members of their communities in terms of education, well-being, or mental health, yet "all were deeply religious, believing their actions sanctioned by the divinely revealed religion of Islam" [44, p. 1537]. Deeply rooted religious beliefs made these people an easy target for leaders of religious extremism, who presented themselves as the ambassadors of God. As shown in the studies, coping with the threat of death consumes a significant of psychic energy [28]. One can assume that the explicit belief that the leaders' orders are sanctioned by God can generate the amount of psychic energy sufficient for committing an act of heroism during a war or an act of suicidal terror.

It is possible that the psychic energy generated by the implicit belief in the divine origins of the people of power, which is a characteristic feature of many modern rational adults, can be used for mind control as well. To a considerable extent, our impulse to go along with suggestions by political leaders, to conform and obey might be powered by the subconscious belief that the commands come from entities blessed by 'divine powers'. For example, many victims of Stalin's repressions, wrongly accused of crimes they had not committed, died with words of blessing to Stalin. Similarly, in Nazi Germany many people collaborated with the regime willingly, even in circumstances under which there was no prospect of retaliation for disobedience [45]. On a smaller scale, the manipulation can have less massive but no less tragic consequences: thus, the unwillingness of aircraft crewmembers to challenge the captain's obvious errors has been named as a cause of plane crashes [46].

All and all, psychic energy generated by our magical experiences and explicit or implicit beliefs in magic is a hidden resource that can be reached and used in education, politics, economics, and medicine.

3.8 Taming the Magic: Conclusion

Research reviewed in this chapter suggests that magical thinking, magical experience and the belief in magic are manifestations of living consciousness present throughout history, cultures, and the lifespan. Unlike magical thinking and magical experience, which remain a conscious practice throughout the lifespan, the belief in magic in most adult educated individuals is ousted into the domain of unconscious. Research on living consciousness links together phenomena that thus far have been studied in separation: magical beliefs in ancient and modern cultures, children's magical beliefs, superstitions in adults, religious beliefs, and the use of magical effects in the entertainment industry. Understanding the laws of living consciousness may provide practitioners with useful insights in the domains of cognitive development, education, and medicine.

The laws of magical thinking can explain manipulations of people's living consciousness. Revealing the psychological basis of such manipulations can have substantial practical consequences. It is one thing to believe that you comply with political or economic requests on rational grounds and quite another to know that you do this due to your hidden belief in magical powers. Understanding the mechanisms of living consciousness can help people to be more cautious and rational in their decision-making.

In Part 3 of this book, we will consider magical manipulation of the living consciousness of individuals, groups, and societies in more detail. Before we do this, however, we need to put living consciousness of a person in broader psychological and historical perspectives.

References

1 https://en.wikipedia.org/wiki/Law_of_thought#Schopenhauer
2 Rozin, P., Millman, L., & Nemeroff, C. (1986). Operation of the laws of sympathetic magic in disgust and other domains. *Journal of Personality and Social Psychology, 4*, 703–712.
3 Dalton, R. (2003, September 4). Lion man takes pride of place as oldest statue. *Nature, 425*, 7. https://doi.org/10.1038/425007a
4 Nemeroff, C., & Rozin, P. (2000). The making of the magical mind: The nature and function of sympathetic magical thinking. In K. S. Rosengren, C. N. Johnson, & P. L. Harris (Eds.), *Imagining the impossible: Magical, scientific, and religious thinking in children* (pp. 1–34). Cambridge University Press.
5 Woolley, J. D. (1997). Thinking about fantasy: Are children fundamentally different thinkers and believers from adults? *Child Development, 98*, 991–1011.
6 Subbotsky, E. (2004). Magical thinking in judgments of causation: Can anomalous phenomena affect ontological causal beliefs in children and adults? *British Journal of Developmental Psychology, 22*, 123–152.
7 Frazer, J. G. (1922). *The Golden Bough*. Macmillan.
8 Lévy-Bruhl, L. (1923). *The primitive mentality* (English translation). Alcan.
9 Bühler, C. (1937). *From birth to maturity: An outline of the psychological development of the child* (Esther, Menaker W., Trans.). Kegan Paul, Trench, Trubner.
10 Piaget, J. (1971). *The child's conception of the world*. Routledge.
11 Subbotsky, E. V. (1985). Preschool children's perception of unusual phenomena. *Soviet Psychology, 23*, 91–114. https://www.researchgate.net/publication/232591513
12 Harris, P. L., Brown, E., Marriott, C., Whittal, S., & Harmer, S. (1991). Monsters, ghosts, and witches: Testing the limits of the fantasy-reality distinction in young children. *British Journal of Developmental Psychology, 9*, 105–123.
13 Woolley, J. D., Boerger, E. A., & Markman, A. B. (2004). A visit from the Candy Witch: Factors influencing children's belief in a novel fantastic entity. *Developmental Science, 7*, 456–468.

14 Subbotsky, E. (2005). The permanence of mental objects: Testing magical thinking on perceived and imaginary realities. *Developmental Psychology, 41*, 301–318.
15 Cameron, J., & Pierce, W. D. (1994). Reinforcement, reward, and intrinsic motivation: A meta-analysis. *Review of Educational Research, 64*, 363–423.
16 Lepper, M. R., Corpus, J. H., & Iyengar, S. S. (2005). Intrinsic and extrinsic motivational orientations in the classroom: Age differences and academic correlates. *Journal of Educational Psychology, 97*, 184–196.
17 Harlow, H. F. (1958). The nature of love. *American Psychologist, 13*(12), 673–685.
18 Mendel, G. (1965). Children's preferences for differing degrees of novelty. *Child Development, 36*(2), 453–465.
19 Berlyne, D. E. (1950). Novelty and curiosity as determinants of exploratory behaviour. *British Journal of Psychology, 41*, 68–80.
20 Henderson, B., & Moore, S. G. (1980). Children's responses to objects differing in novelty in relation to level of curiosity and adult behaviour. *Child Development, 51*(2), 457–465.
21 Comerford, B., & Witryol, S. L. (1993). Information metrics for novelty level preference of first-and fifth-grade children. *The Journal of Genetic Psychology, 154*(2), 155–165.
22 Pliner, P., Pelchat, M., & Grabski, M. (1993). Reduction of neophobia in humans by exposure to novel foods. *Appetite, 20*, 111–123.
23 Rozin, P., Haidt, J., & McCauley, C. R. (1993). Disgust. In M. Lewis & J. Haviland (Eds.), *Handbook of Emotions* (pp. 575–594). Guilford.
24 Jahoda, G. (1969). *The psychology of superstition*. Penguin.
25 Zusne, L., & Jones, W. H. (1982). *Anomalistic psychology*. Erlbaum.
26 Subbotsky, E. (2010). Curiosity and exploratory behaviour towards possible and impossible events in children and adults. *British Journal of Psychology, 101*, 481–501.
27 Gailliot, M. T., Baumeister, R. F., DeWall, C. N., Maner, J. K., Plant, E. A., Tice, D. M., Brewer, L. E., & Schmeichel, B. J. (2007). Self-control relies on glucose as a limited energy source: Willpower is more than a metaphor. *Journal of Personality and Social Psychology, 92*, 325–336.
28 Gailliot, M. T., Baumeister, R. F., & Schmeichel, B. J. (2006). Self-regulatory processes defend against the threat of death: Effects of self-control depletion and trait self-control on thoughts and fears of dying. *Journal of Personality and Social Psychology, 91*, 49–62.
29 Blake, T. M., Varnhagen, C. K., & Parent, M. B. (2001). Emotionally arousing pictures increase blood glucose levels and enhance recall. *Neurobiology of Learning and Memory, 75*, 262–273.

30 Subbotsky, E., Hysted, C., & Jones, N. (2009). Watching films with magical content facilitates creativity in children. *Perceptual and Motor Skills, 111*, 261–277.
31 Subbotsky, E., & Slater, E. (2011). Children's discrimination of fantastic vs realistic visual displays after watching a film with magical content. *Psychological Reports, 112*, 603–609.
32 Subbotsky, E., & Matthews, J. (2011). Magical thinking and memory: Distinctiveness effect for TV advertisements with magical content. *Psychological Reports, 109*, 1–11.
33 Frasier J. (1993, April 14–18). Magic in a classroom: Using conjuring to teach selectivity and general semantics. Paper presented at the Annual Meeting of the Central States Communication Association, Lexington, KY.
34 McCall, B. (2008, September 12) Teach magic lessons in school, says professor. Guardian. http://www.guardian.co.uk/education/2008/sep/12/secondaryschools
35 Wiseman, R., & Watt, C. (2018). Achieving the impossible: A review of magic-based interventions and their effects on wellbeing. Brain and Cognition. https://peerj.com/articles/6081/
36 Wiseman, R., Wiles, A., & Watt, C. (2021). Conjuring up creativity: The effect of performing magic tricks on divergent thinking. *Journal of Peer, 9*(2), e11289.
37 http://www.ebay.fr/itm/Jean-Effel-The-Creation-of-the-World-book-cartoon-caricature-226-pg-1963/272365953720
38 https://map.gsfc.nasa.gov/universe/
39 Subbotsky, E. (2015). I believe because it is absurd: The magic of religion. *Psychologist, 6*, 76–115.
40 Ivanov, U. G. (2003). *Kamikaze: Suicidal pilots. Japanese self-sacrifice during the war in the Pacific* (2nd ed.). Rusich.
41 Barker, L. (2002). *Psychology*. Pearson Education.
42 Westen, D. (2002). *Psychology: Brain, behaviour, and culture*. John Wiley & Sons.
43 Atran, S., Axelrod, R., & Davis, R. (2007). Sacred barriers to conflict resolution. *Science, 317*, 1039–1040.
44 Atran, S. (2003). Genesis of suicidal terrorism. *Science, 2999*, 1534–1539.
45 Fromm, E. (1961). *The fear of freedom*. Routledge.
46 Kanki, B., & Foushee, H. C. (1990). Crew factors in the aerospace workplace. In S. Oskamp & S. Spacepan (Eds.), *People's reactions to technology* (pp. 18–31). Sage.

PART II

Living Consciousness and the Universe

CHAPTER 4

Man in the Mirror of Nature: Science and Living Consciousness

4.1 Problem: The Cosmic Solitude

According to the 'Copernican principle' accepted in modern astrophysics, humans are not privileged observers of the universe; instead, humanity lives among the vast expanses of cold and indifferent space on a tiny planet doomed to death in the distant future [1]. Nature and the cosmos existed for billions of years before the emergence of people and will exist for billions of years after their disappearance. The so-called 'cosmological principle', recently confirmed by the discovery of the cosmic microwave background radiation [2], states that in the universe matter is equally distributed and isotropic, meaning that at any of the arbitrary taken point in the physical universe a person will see the same picture and find the same physical laws [3]. The attempts of scientists to smoothen the 'Copernican mediocrity' [4] by pointing out that the universe might be not as isotropic as it seemed [5] and that Earth is a unique planet situated in the extremely rare 'goldilocks zone' [6] don't save the humanity from the perspective of inevitable death in the future and don't answer the question of why the intelligent life exists at all. The ecological degradation of our planet because of human expansion might even provoke one to think that humanity is a kind of mould—a parasite sucking juices of the primordial Mother Earth [7].

But what if we assume that all these billions of years before the emergence of intelligent observers and all the infinite number of light years

E. Subbotsky, *The Magic of Living Consciousness*,
https://doi.org/10.1007/978-3-031-54708-9_4

beyond the horizon of the universe is nothing more than an illusion? Powerful, difficult to overcome, but still an illusion? What if in rejecting the belief that the Earth is the centre of the cosmos Nicolaus Copernicus was right about the planets, but not about a person, and the person is indeed a centre of the universe? Today such an assumption may seem somewhat unexpected, yet there are 'special facts' in science that make one take a closer look at this assumption.

One of these special facts is the so-called 'fine tuning' of the universal cosmological constants. The 'fine tuning' means that if the fundamental cosmological parameters of the universe, such as the ratio between masses of the electron, proton, and neutron, or the famous Planck's constant, were at least a tiny fraction of a percent different, the existence of life and intelligent observers would be impossible. Based on the fact of the 'fine tuning', the strong anthropic principle states that the universe must necessarily have properties that allow the emergence of intelligent life [8]. This means that physical properties of the universe and human consciousness are connected. The famous American physicist John Wheeler wrote "all things physical are information-theoretic in origin and … this is a participatory universe" [9, p.5]. Wheeler's statement resonates with the old philosophical dilemma: "Does a tree falling in a dense forest make a crack, on the condition that there is no one nearby at that moment who would hear this crack?" If we say that it does, we are already mentally present next to the tree and violate the condition. We must admit that without intelligent observers there would be neither the Earth nor the universe. Through the human observer, through his or her living consciousness the universe passes from the state of Nothing to the state of Being.

Another 'special fact' is that in the quantum world the essence of a quantum object in motion cannot be established before it is 'stopped' and fixed by a measuring device. For example, it is impossible to say whether this object is a particle that has a certain position in space or a wave that hasn't. What this quantum object will become when it is fixed by a measuring device is determined by the properties of the measuring device and the procedure of measurement [10], and any man-made measuring device is an artificial extension of our natural 'measuring devices'—eyes, ears, and living consciousness. The phenomenon of 'quantum entanglement' recently recognized as proven (Nobel Prize in Physics for the year 2022) showed that a measurement of one of the pair of 'entangled' quantum objects can instantly affect the state of the 'partner', even if the partner is at the border of the visible physical universe [11, 12]. If, as follows from the 'Copernican principle', a person were isolated from the rest of the cosmos on an insignificant planet Earth, such a connection between a person's living consciousness and the cosmos, both in the micro- and the mega-worlds, would be impossible. So, in light of the data of modern physics, the 'Copernicus principle' does not work for a person, and the person's living consciousness indeed affects the state of the physical universe.

Given the 'special facts' we are faced with the choice of either admitting that the physical universe was 'programmed' by God in such a way that the emergence of living consciousness was inevitable, or assuming that there is an innumerable number of parallel universes in one of which fundamental cosmological constants were lucky to coincide in such a unique way that there was a possibility of the emergence of living consciousness. The first choice requires faith, but not all scientists can genuinely believe in God. The second possibility is more in peace with science, but it contradicts the principle of 'participatory universe'. Suppose that a unique combination of fundamental cosmological constants in our universe arose by chance, but intelligent observers could not emerge instantly and billions of years had to pass before living consciousness appeared during evolution. How then could the physical universe exist before the appearance of living consciousness, if the existence of living consciousness is a necessary condition for the emergence of the universe? Since both options are fundamentally inaccessible to experimental verification, one must look for an alternative explanation of the relationships between living consciousness and the universe.

The problem we are facing is therefore how, without resorting to the hypotheses of creation or the postulation of an innumerable number of parallel universes, to explain that: (1) the universe existed billions of years before the emergence of living consciousness given that the existence of living consciousness is a necessary condition for the existence of the universe, and (2) the universal cosmological constants are tuned to each other in such a way as if the emergence of living consciousness was pre-planned?

Let's see if the properties of living consciousness can help to answer these questions.

4.2 Personal Universe

As already mentioned, some scientists believe that the 'fine tuning' of universal cosmological constants is a random coincidence that occurred in one of an infinite number of alternative universes. But such a view presents the universe as an object that exists separately and outside of intelligent observers, like stones, mountains, and galaxies. Although it is possible for a person to view stones, galaxies, and blackholes as separate from himself or herself, the person cannot separate himself or herself from entities that he or she is a part of, such as the universe, nature, or evolution. When we look at the physical universe from the 3rd person's perspective, unresolvable paradoxes appear, such as the pre-conditioned 'fine tuning' of universal cosmological constants necessary for the emergence of the intelligent observer, or the Big Bang, when the physical universe arises virtually from nothing [13]. These paradoxes suggest that the view of the universe as the object outside the person's living consciousness has its limits, and that along with viewing the universe from the 3rd person's perspective the *reversed perspective* is possible—the perspective in which a person looks at objects within which he or she is located as if these objects are a part of his or her own living consciousness (see Fig. 4.1).

Indeed, in the world that we perceive a person cannot go beyond the limits of the physical universe or his or her own body simply because the person has nowhere to go. Any person, including a scientist, is inside the physical universe, which appears to him or her as an external world but can also be viewed as a part of his or her own living consciousness. As argued in Chapter 1, the role of an intelligent observer is performed by the person's internal Self. In contrast to the person's external Self (his or her body and perceptions), the internal Self is a controlling subject, which the person cannot observe as a phenomenon but can sense intuitively as his or her 'I' (see [14] for more of the differences between external and internal Selves). To deal with the physical universe, the scientist's internal Self operates with subjective experiences, such as perceptions, knowledge, and thinking, the logical part of which the scientist learned from books, and the intuitive and creative part is given to him or her from birth. To these subjective experiences and innate abilities, one can add imagination, memory, and faith. And all these subjective experiences and abilities, operated by the internal Self, are nothing more than different forms of the same thing—the living consciousness of a particular person. In this

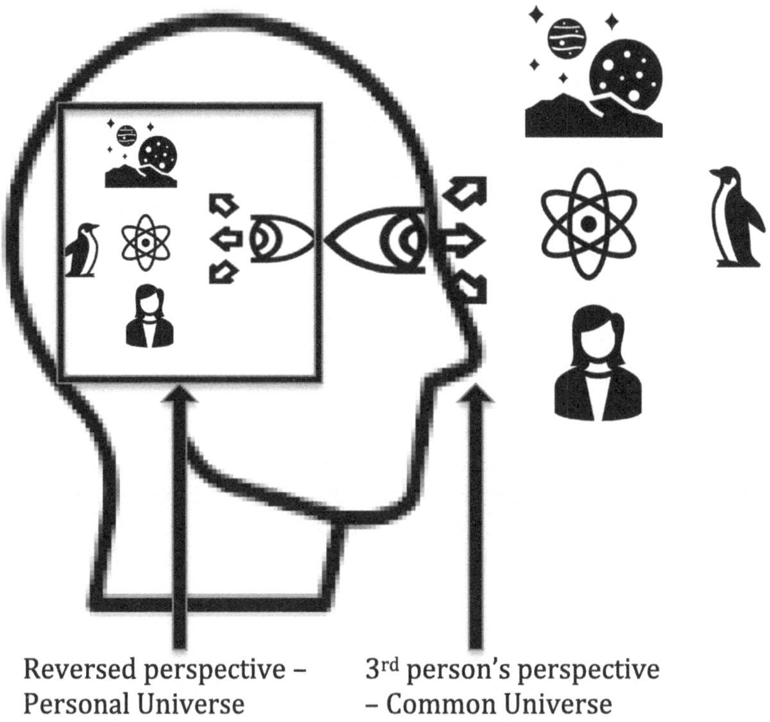

Fig. 4.1 Common Universe and Personal Universe

chapter, viewing the universe from the perspective as if the universe was a part of the observer's living consciousness will be referred to as the perspective of the personal universe, or simply *Personal Universe*.

Viewed as a part of Personal Universe, the 'fine tuning' of the fundamental cosmological constants turns from a cosmological mystery into a necessary condition. This follows from the fact that the fundamental cosmological constants are knowledge obtained by scientists through the perception of phenomena available in their Personal Universes, and the application of mathematics to these perceptions. But mathematics is knowledge as well, acquired through perception and thinking. For example, mathematicians Euclid or Lobachevsky had to first see parallel lines on a plane or on a sphere, and only then they could formulate axioms about parallel lines. This means that the fundamental cosmological constants in the Personal Universe must necessarily be in such harmony with each other that allows the existence of the intelligent observer, simply because if there were no such harmony there would be neither the intelligent observer nor the cosmological constants. At the same time, the Personal Universe does not need the Big Bang hypothesis, since it is attached to a specific person and does not have past and future independent from the person; the past and the future of the Personal Universe are constructed by the person and exist together with the person while he or she lives and thinks.

The 'fine tuning' of the universal cosmological constants can be illustrated by the harmony of various parts of a human organism. One might be surprised by the fact that body organs such as a heart, a liver, and lungs (read: universal cosmological constants) fit together so well only if one forgets that scientists discovered these organs as parts of an already existing whole—the human body (read—the Personal Universe). In other words, the strong anthropic principle looks mysterious only if we assume that the universe existed billions of years without intelligent observers, which from the point of view of Personal Universe is impossible, since Personal Universe cannot exist without its owner—living consciousness of a given individual person. So, when a cosmologist discovers that the universal cosmological constants are tuned in a way that allows the existence of an intelligent observer, the cosmologist sees the constituent components of his or her own living consciousness.

Summing up, from the point of view of Personal Universe, an answer to the problem posed at the beginning of this chapter is as follows: the *physical universe* (e.g., cosmological constants) along with the *psychological universe* (e.g., perceptions, thoughts, and desires) and *social universe* (society and culture), is a part of the Personal Universe (see Table 4.1). The fundamental cosmological constants are components of the physical universe; therefore, these constants necessarily must be 'fitted' to each other, thus making the Personal Universe possible. Importantly, our internal Self can control only certain parts of our psychological universe, such as intentional movements of our body and voluntary mental actions, whereas events in physical and social universes do not obey our internal Self but have laws of their own. These laws are studied by the sciences. Since the existence of our Personal Universe is limited by 'here and now', the past and the future are artificial extensions within our Personal Universe, which make our present meaningful. The past and the future of the Personal Universe are inaccessible to perception. Therefore, we *see* and ***know*** the present, but we ***believe in*** the past and the future. Speaking figuratively, the past and future of our Personal Universe is a 'rope-walker's pole', which gives us the opportunity to maintain balance on the tense string of the present.

The 'fine tuning' of the fundamental cosmological constants is just one necessary consequence of the Personal Universe hypothesis. Another necessary consequence is that a scientist can explore the physical universe only by projecting properties of his or her psychological universe onto the physical and social universes [ibid]. Indeed, science is made by specific people inside their Personal Universes on the basis of their living consciousness, such as knowledge, memory, imagination, and creative thinking. Our psychological universe (sensations, feelings, desires, and thoughts) is given to us from birth, whereas properties of the physical and social universes are established by our internal Self through experience. Sweet and bitter, white and red, quiet and loud, empty and full, soft and hard are projections of our psychological universe onto the outside world, whereas atoms, molecules, quantum objects, and galaxies are intellectual constructions created on the basis of these projections with the help of thinking. We know that clouds are water molecules in the sky, but what

Table 4.1 The structure of Personal Universe

Components *Type of consciousness*	*Psychological universe*	*Physical universe*	*Social universe*
Hard (objectified)		Science generated objects (physical, cosmological, chemical, and biological)	Science generated objects (sociological, economical, ethical, linguistic, philosophical) and social conventions (customs, traditions, moral and social laws)
Soft (living)	The person's 'internal Self', perceptions, thoughts, desires, needs, beliefs, phantasies and subjective experiences (pain, pleasure, dreams and emotions)		Needs, beliefs, collective narratives and 'dreams' of a society

we see are not molecules but clouds, sometimes looking like a person, a bird, or a medieval castle.

So, in order to cognize the physical and social universes we 'objectify' properties of our living consciousness by projecting them onto physical and social universes. It is necessary to emphasize though that this projection doesn't mean creation from scratch. Potentially, properties and laws of physical and social universes are independent of our living consciousness, but they come into existence in our Personal Universe only by being 'dressed' into the 'fabric' of our living consciousness. This means that our living consciousness ***participates*** in the way we see and understand the physical and social universes. Put in other words, our living consciousness is a co-creator of physical and social universes. Moreover, the projection of our living consciousness onto physical and social universes not only participates in the way these universes present themselves to our internal Self, but also allows our internal Self to cognize our psychological universe.

An example of projection of our living consciousness onto the physical universe can be found in biology. The more we study animal behaviour, the more we see it as a reflection of ourselves. First, we discovered the ability to associate effects with causes in ourselves, and then we found similar abilities in rats and fish [15, 16]. We located the ability to empathize with others in ourselves and then found it in mice and birds [17, 18]. When biologists and psychologists study the animal kingdom, they see living consciousness of their own in the mirror of nature. Like those physicists who are surprised by the 'fine tuning' of the fundamental cosmological constants, many biologists interpret the humanoid skills of animals as something completely independent of their own living consciousness; in fact, these scientists 'turn the tables' on cause and effect and present the properties of their psychological universe as a result of the evolution of the animal kingdom [19, 20].

"Are you saying that evolution is not an objective reality but a creation of our mind?" the reader might ask. No, it is not, but the phenomenon of evolution depends on the properties of our living consciousness. From the perspective of Personal Universe, evolution, like other laws of physical universe, is a result of the projection of our psychological universe onto the external reality merged with this reality's own properties. When we see our reflection in a mirror, we discover two things at one time: the property of the mirror to reflect things, and our own image reflected in the mirror. In a similar vein, by projecting certain properties of our living consciousness on biological nature, we discover evolution and at the same time see the properties of our living consciousness in a brighter light.

One of many properties of our living consciousness is 'synthetic creativity'—the ability to put simple things together to create a more complex object in our imagination. Observations show that this ability is innate: young infants assimilate the external environment to fit their needs and then extend this assimilation to form more complex schemata [21], young children create stable artificial configurations from sand using plastic forms, older children construct new structures from the Lego blocks and engineers built complex structures from simpler compounds. Projecting the synthetic creativity of our living consciousness on external reality leads to the realization that synthetic creativity works in the physical universe as well. We call this property evolution.

Thus, having observed that artificial selection of animals brings about new breeds of cattle and dogs, Charles Darwin created his theory of natural selection, in which the synthetic creativity was displayed not by an intelligent being but by 'intellectually blind' natural forces, such as the living organisms' instinct to survive, the natural tendency of these organisms to change, replicate and save the spontaneously acquired properties in subsequent generations, coupled with the changing conditions of the environment [22]. By doing this Darwin projected the synthetic creativity of living consciousness onto the physical universe in the domain of biology and discovered a new phenomenon in nature. Extending such a projection, some modern theorists see the synthetic creativity working in the domain of non-living things leading to the increasing complexity of atoms, molecules, and chemical compounds [23]. If such a projection succeeds, we create the 'evolutionary ladder', and if it fails, then we conclude that this particular property or ability is specific only to human consciousness. For example, when we observe how our dog sleeps and tries to 'catch a fly' by clapping its teeth, we conclude that dreaming is a common feature of the psyche of dogs and humans, and when we notice that the dog cannot speak English, we conclude that spoken language is an exclusive property of our living consciousness. We usually say that the first function is 'evolutionarily programmed' and the second is not.

But the projection of living consciousness on nature harbours a danger of reversing the relationship between cause and effect; as a result, evolutionary thinking can turn from a useful mechanism of understanding nature and self-cognition into an explanatory trap. Thus, projecting synthetic creativity of the living consciousness into non-organic matter may turn into interpreting our living consciousness as a *final stage of the evolution* of synthetic creativity working independently in physical universe. In a similar vein, when we observe that behaviour of animals is determined by instincts and reflexes, we might conclude that we too are nothing more than 'biological machines' and our sense of free will is an illusion. There is only one way out of this trap—to realize that the illegitimate reversal of cause and effect has occurred and not all abilities of our living consciousness can be reflected in the mirror of evolution, just as we cannot see our thoughts in a real physical mirror. Such properties of living consciousness as free will, sense of beauty, faith in gods, sense of moral duty, and some others are typically human, and attempts to project these properties into the animal world is doomed to fail.

But if reason, morality, and faith can't be found in the animal kingdom, then perhaps these properties can be replicated in the world of machines. Indeed, the rapid development of artificial intelligence (AI) in the XXth century gave rise to the hopes that such replication is possible. Today machines can 'recognize' people's faces, 'make' logical inferences at a speed inaccessible to humans, and 'win' chess games against world champions. In other words, the AI (sometimes called the nature-plus) becomes a new 'mirror of Narcissus' [24], and people who have unlimited faith in science are increasingly falling in love with their own reflection in this man-made mirror. But there is a price to pay for this passion: if the mythical character—the handsome Narcissus—due to an unrequited passion for his own reflection in the water died and turned into a flower, some scientists today, at least in theory, turn their living consciousness into a product of machines consisting of molecules, neurons, and bits of information.

4.3 The Revolt of Machines

The creators of the philosophical foundations of modern science—Rene Descartes, Baruch Spinoza, and Immanuel Kant—considered living consciousness as a gift of God. Today, in the scientific literature there is the widely spread view that phenomena of living consciousness—perceptions, feelings, and images—are the product of information processes in the brain. In some sense, this view can be presented as a revolt of machines.

Indeed, consider a typical interaction between a person and a machine, from the simplest (e.g., a spoon) to the most complex (e.g., a computer).

Usually, we believe that in this interaction the person is a controlling subject and the machine is an instrument, despite the machine can affect the person's living consciousness by forcing the person to acquire new skills. It is possible, however, to begin seeing the hierarchy in a different way and believe that the machines do not simply influence living consciousness but produce it. As a result of this upheaval in relations between living consciousness and machines, a cause (the living consciousness) becomes an effect, and an effect (the machine) becomes a cause.

The revolt of machines is a popular theme in science fiction films. Machines manipulate people's living consciousness while subjugating their bodies to use them as a source of energy (The Matrix) [25], they acquire control of the person's body (Upgrade) [26] or absorb the person's living consciousness entirely (Lawnmower Man) [27]. While such a dystopian future seems a distant prospect, it is already happening, if by machines we mean entities of objectified consciousness—*science generated objects (SGOs)*. Among the giant pool of SGOs, there are three that have spearheaded the toppling of living consciousness from its pedestal: electricity, the brain, and information.

The production of SGOs was a great achievement of human genius, but it harboured a hidden danger. Humans faced this danger when because of the spectacular advances in science over the past 4 centuries the Copernican revolution in cosmology was replicated in psychology and some scientists began to view their living consciousness as the result of the functioning of the SGOs.

4.4 Neuroscience in the Narcissistic Mirror

Today many scientists believe that the brain is a biological computer, consisting of millions of 'chips'—neurons—that conduct electrical impulses. Nerve impulses transmit information from the outside world and create what we take for phenomena of living consciousness. This makes the living consciousness an epiphenomenon—a shadow that accompanies nervous processes in the brain but doesn't have causal powers in the physical universe [28][29][30]. Moreover, the concept of information has been put in the foundation of the physical universe as a whole: in the 'theory of digital ontology' the physical universe is understood as a simulation on the screen of a 'cosmic supercomputer' [31] (see Chapter 5 for the elaboration on that).

The revolt of machines led to the situation when living consciousness, which has a leading role in the creation of the SGOs, began to be viewed as a product of the SGO's, and the person from God's favourite creation turned either into an evolutionarily evolved 'vehicle' for carrying information embedded in the genes, or into a 'character' of the cosmic computer game. The revolt of machines culminated in theories of some distinguished representatives of neuroscience.

Neuroscience explores correlational links between phenomena of living consciousness (e.g., your perceived or imagined phenomenon of a tree) and objectified consciousness (e.g., a specific group of neurons firing at that moment in your brain). However, some neuroscientists replace correlations with determination by considering brain processes as a cause of living consciousness. As the Nobel Prize-winning neuroscientist put it, "the last frontier of the biological sciences—their ultimate challenge—is to understand the biological basis of consciousness and the mental processes by which we perceive, act, learn and remember" [32, p. 5]. Pushing this thesis even further, another distinguished neuroscientist writes: "You, your joys and your sorrows, your memories and your ambitions, your sense of personal identity and free will, are in fact no more than the behaviour of a vast assembly of nerve cells and their associated molecules" [33, p. 3].

If brought to its logical end, this line of thought opens the possibility of reducing the phenomenal variety of living consciousness to predictable and controllable patterns of electrical activity in the brain. If this goal were achieved, a neuroscientist could create a phenomenon of living consciousness in a patient's mind (for example, an image of a person sitting in the garden) by turning certain neurons on. Since the created phenomenon had already been present in the mind of the neuroscientist who manipulated the patient's brain, these manipulations would be a *forced simulation* of the neuroscientist's living consciousness in the patient's mind. The alternative to the forced simulation would be to simply ask the patient to imagine a person sitting in the garden. But for the ideologists of the forced simulation this isn't enough, they want to be able to control the patient's living consciousness by directly manipulating the processes in the patient's brain.

The person who gave rise to the hope that the forced simulation is possible was the Canadian neurosurgeon Wilder Penfield. In the 1950s Penfield was looking for a way to relieve seizures in patients suffering from particularly severe cases of epilepsy. He used to open the patient's temporal lobe and stimulate the cortex with a low-voltage electric current. Penfield registered a new phenomenon, which he called dual consciousness. Patients under local anaesthesia and in full living consciousness reported two parallel but separate streams of subjective experiences: One artificially created by the electrical stimulation of the cerebral cortex, and the other caused by perceptions of the current situation in the operating theatre. At the same time, the patients could accurately distinguish the artificially induced stream of living consciousness from the real one [34].

However, the dual consciousness is only an intermediate stage on the way to forced simulation, since electrical stimulation of the cerebral cortex can only evoke a certain unpredictable content of living consciousness, whereas the forced simulation is supposed to be an exact copy of the content that had been assigned by the neuroscientist in advance. The problem of implementing the forced simulation is complicated by the fact that the results of electrical stimulation of the brain can only be evaluated by the patient's conscious report. The way out of this difficulty would be if the neuroscientist stimulated the neurons in his or her own brain to evoke certain subjective images that he or she had assigned in advance. For example, the neuroscientist had assigned to see a squirrel eating a walnut, and by stimulating his or her brain the neuroscientist saw the image of the squirrel eating the walnut.

However, it is complicated to achieve a strict match between the excited neural patterns and the subjective images that correlate with them. One of the reasons is the complexity of the neural process, carried out by thousands of molecules within a single neuron, which, in turn, acts in concert with a network of tens of thousands of neurons, which are only a fraction of about 90 billion neurons in a single human brain [35]. Another reason is that establishing the correspondence between neural activity and living consciousness is fundamentally different from converting a digital form of information into an analogue one. When electronic impulses are converted into light signals on a computer screen, two physical phenomena interact—the electrical impulses and the light signals [36]. Similarly, when a DNA molecule determines the structure of proteins that make up our body, physical phenomena—DNA and proteins—are connected by a causal sequence. However, when a neuroscientist wants to create a certain image of living consciousness in his or her mind by stimulating the neurons in the brain, the neuroscientist must overcome the ontological barrier—turn phenomena of objectified consciousness into the phenomena of living consciousness. But the main obstacle to such a transformation is that such a transformation contains a *logical circle*.

A logical circle is a fallacy in which a statement is inferred from itself, usually through several intermediate statements. Specifically, to initialize certain neurons in his or her brain with the aim of producing a premeditated subjective image in his or her living consciousness, the neuroscientist must already have the desired image available; as a result, the effect (subjective image in the neuroscientist's living consciousness) generates its own cause (excitation of certain neurons in the brain). This mental experiment reveals that viewing subjective images of living consciousness as generated by processes in the brain involves a logical error: the *correlation* between the excitation of the cerebral cortex and the subjective report of the patient is taken for a *causal relationship*. What is causing this error?

The cause may be hidden in the difference between existential weights of the SGOs and phenomena of living consciousness. The *existential weight* of an entity is a person's belief in the significance of this entity for life and practice. The more noticeable and significant for a person's life the entity is, the greater the entity's existential weight. The existential weight of a cause of some phenomenon can be much less than the existential weight of an effect of this cause. For example, an avalanche (the effect) may have a stronger existential weight than its cause—the shift of a small pebble on top of the mountain, and a dangerous disease (the effect) may be caused by a virus invisible to the naked eye.

The successes of science over the past four centuries have led to the fact that the existential weight of the subjective phenomena of living consciousness has become much smaller than the existential weight of the phenomena of objectified consciousness such as SGOs. Indeed, subjective phenomena of living consciousness are individual, unique, unstable, and one-sided. On the contrary, SGOs are inter-individual, reproducible, stable, and comprehensive. As an example, take our perception of a house and its building plan. Our perception (subjective phenomenon of living consciousness) of the house changes depending on the angle of view, we see only a visible part of the house and cannot convey our subjective image to others. The building plan (SGO) is stable, transferable, and comprehensive.

All and all, the fabric of living consciousness cannot be inferred from neural activity, because neural activity is a SGO that implicates living consciousness of scientists who created it. This doesn't mean that neuroscience cannot achieve significant advances by providing knowledge about flexible correlations between subjective phenomena of living consciousness and the activity of certain brain regions. But these advances, useful and interesting as they are, cannot provide neuroscience with the biological explanation of living consciousness.

Now let's look at how subjective phenomena of living consciousness gradually turned into SGOs, in relation to the key concepts - electricity, brain and information.

4.5 'Digitalization' of Living Consciousness

Let's start with *electricity*. Around 600 BC the Greeks noticed that rubbing petrified tree resin (amber) with animal fur caused the resin to attract dry grass. Around 1600 AD, the English physicist William Gilbert used the Latin word 'electricus' to describe this phenomenon of living consciousness. In 1752, the American polymath and politician Benjamin Franklin conducted his experiment with a kite, a key, and a thunderstorm, which proved that lightning and tiny electrical sparks were one and the same. Subsequent experiments on this phenomenon by Alessandro Volta, Luigi Galvani, André-Marie Ampère, George Ohm, Michael Faraday, and other scientists led to the appearance of the SGO we know as electricity. So, when you turn on the lights in your apartment and read a hefty volume on the physics of electricity, remember that it all started with the modest phenomenon of living consciousness—a perception of the effect of rubbing a piece of amber with a bunch of animal fur.

Now let's take *the brain*. The ancients' views on the brain's functions varied. The Greek scientist Hippocrates considered the brain to be a seat of consciousness, while the Greek philosopher Aristotle thought that the brain was a mechanism for cooling the blood. Approximately two millennia later, the French philosopher René Descartes proposed that the pineal gland in the brain is the place where living consciousness (spirit) relates to the body. In 1780, the Italian physician Luigi Galvani discovered that touching an electric spark to the leg muscles of dead frogs caused the leg to twitch. At the end of the nineteenth century, the Spanish anatomist Santiago Ramon y Cajal saw under a microscope a section of the substance of the chicken cerebellum, consisting of units looking like a squid with many tentacles, which he took for a functional unit of the brain. These units, called neurons, were connected to each other by a network of joints that the British physiologist Charles Sherrington called synapses. More recent discoveries in the field of neuroscience have shown that our thoughts and actions are correlated with the electrical activation of certain areas in the spherical lump of neurons that sits in a human skull. This more than 2 millennia-long chain of phenomena of living consciousness led to the creation of a new SGO: the brain. Thus, a lump of greyish matter in the skull came to be viewed as the seat of living consciousness.

One more Cinderella story was the invention of a *unit of information*—a bit. The story begins with the observation by ancient philosophers that things in the world are structured as opposites: empty and full, high and low, up and down, on and off. Encoding data in the form of discrete bits was invented in 1732 by the French textile workers Basile Bouchon and Jean-Baptiste Falcon, who used the opposition 'hole or no hole' to make a perforated paper to partially automate textile production. Computer manufacturers such as IBM have copied this method. In early computers, a card or tape carried many hole positions; each position could be pierced or not pierced, thus transmitting one bit of information. A byte is a larger unit of eight bits. A byte can be used to encode a single character of text or a number. For example, the character A can be encoded as 00,000,001, the character B as 00,000,011, the number 1 as 10,000,000, the number 2 as 10,000,001, and so on. With more bytes, almost anything can be encoded: letters, numbers, colours, shapes, sounds, and musical notes. Combined with electricity and technical devices such as semiconductor chips, digital code can create technical miracles—all those beautiful pictures and sounds on our TV screens, computers, and smartphones that simulate the phenomenal splendour of living consciousness with ever-increasing perfection. So, a group of phenomena of living consciousness—the observation of opposites in the perceived world, the strip of perforated paper in a loom and creative thinking that put these phenomena together—turned into the SGO known to us as information.

So, from the perspective of Personal Universe, SGOs (electricity, brain, and information) are products rather than causes of living consciousness. As a result of the Copernican existential upheaval in the relationship between living consciousness and the brain there emerged the idea that a subjective phenomenon of living consciousness (e.g., our perception of a room we are sitting in) is a string of information units produced by the brain. This idea of digitalization of living consciousness made it possible to simulate living consciousness in the form of Artificial Intelligence (AI). Unlike the forced simulation of living consciousness via tampering with the brain, the digital simulation is a version of objectified consciousness and essentially a form of artificial language (see Chapter 6 for more on symbolic and digital languages). Today the AI, or 'nature-plus', has become a new mirror of Narcissus, which reflects our living consciousness. Sadly, for many scientists, their living consciousness fell in love with its reflection in the mirror and ceased to take notice of itself.

4.6 Science and Living Consciousness

The task of studying living consciousness is to overcome this existential upheaval and study the properties of living consciousness as the primary reality that gave rise to the SGOs [37]. From the perspective of Personal Universe, the main role of living consciousness is 'pulling' the physical universe from the state of Nothing into the state of Being, and transformation of Chaos into Order [38]. The same applies to the social universe—the world of other people and culture. As studies of infants' cognitive abilities showed, for a newborn child the primary Chaos is already partly ordered. Even in the womb, a child can distinguish the voice of his or her mother from voices of strangers [39]. In the first weeks of life, a child prefers to follow a person's face than to follow figures in which facial features are randomly scattered [40]. In contrast, for a scientist working at the edge of the unknown the Chaos is presented as a bunch of unrelated phenomena that 'speaks' to the scientist's intuition. Like the physical universe, the social universe enters Personal Universe of the scientist through his or her living consciousness.

All and all, the role of science in human life, for all its importance, is auxiliary, *epistemological*, whereas the role of living consciousness is basic, *ontological*. The products of science and culture represent 'hard' objectified consciousness, which is structured and common to all people; in contrast, the psychological universe contains 'soft' living consciousness that doesn't have a rigid structure (see Table 4.1). As indicated earlier, in any act of living consciousness imagination, memory, thinking, will and faith are involved simultaneously. But if living consciousness doesn't have a solid structure, the question arises of how it can be studied.

4.7 Studying Living Consciousness

As already noted, classical scientific knowledge is made within the 3rd person's perspective. This perspective implies that the subject matter of research is independent of both the observer conducting the research and measuring instruments employed in the process. For example, the upward buoyant force that is acting on a body immersed in a fluid is equal to the weight of the fluid that the body displaces, independently of whether the body is a person, a log, or a stone, or whether the liquid is water, oil, or mercury.

Following science, modern cognitive psychology considers living consciousness as a subject matter; as a result, it explores the 'hard' automated fabric of living consciousness—skills in the field of perception, thinking and memory. Thus, when driving a car, we concentrate our attention on watching the road and don't think about how to change the gears, which we change automatically. Such automatized, unconsciously controlled part of living consciousness is the main subject matter of cognitive psychology. For example, if we want to study how many elements (numbers or letters) from a group of 10 elements a person can repeat immediately after the elements are displayed on the computer screen for a short time and then removed, we test a new group of participants every day for 10 days. The results will show that although the number of digits remembered varies from person to person, the average number that people remember on each day of testing will be around 7 (plus minus 2) and this number will remain approximately constant throughout the entire testing period, given that our method and selection of participants remain the same [41].

On the contrary, the study of living consciousness involves the observer effect, like studies in physics, where the consciousness of the experimenter and the measuring instruments can influence the object of study. An example taken from physics is seeing a non-luminous quantum object, which requires light to be reflected by the object, which changes the object of the observation. Similarly, measured effects of living consciousness can change significantly over time, even if the experimental conditions remain constant. The fact that such effects do not obey the laws of nature explains why these effects appear only under specially created conditions, are strictly individual (under the same conditions, some subjects show these effects and others don't), and cannot be fully reproduced. What tasks can be set in the study of living consciousness?

First, it is necessary to understand the laws that govern the functioning of living consciousness. A new area of psychology deals with this problem—the study of magical thinking [42, 43]. For example, in the domain of memory, living consciousness in the form of magical thinking is involved in the process of inventing a mnemonic device—a pattern of letters and ideas that assists in remembering something. Usually, the invented pattern doesn't have a causal or logical relation to the word to be remembered but is linked to this word on the basis of magical participation (see Chapter 3 for the elaboration on that). Thus, when they teach children the rhyme Richard Of York Gave Battle In Vain for remembering the colours of the spectrum (red, orange, yellow, green, blue, indigo, violet) [44], the child's living memory connects the rhyme with the colours of the rainbow via participation based on a common element—the first letters of the words.

Second, it is important to investigate how our internal Self builds physical space, time, and causality from space, time, and causality of living consciousness, and why our internal Self is not limited only to its egotistic needs but because of the mere belief in the existence of other people's Personal Universes creates motives for disinterested moral behaviour—honesty, justice, and altruism [14].

Third, it is fundamentally important to find out whether it is possible for our internal Self to go beyond the limits of our Personal Universe, no matter how paradoxical this task might look. Specifically, is the direct (not mediated by language and behaviour) contact of our living consciousness with the living consciousness of another person (telepathy) possible? Can our living consciousness (thought or desire) directly affect matter (psychokinesis)? Is it possible for a person to see the future not predictable from available knowledge (remote viewing)?

Studies of the so-called anomalous phenomena of the mind show that such effects are possible. Being initially sceptical regarding the possibility of such phenomena, I conducted several studies in this area, independently and in collaboration. I concluded that the phenomena of living consciousness, such as the direct influence of the experimenter's consciousness on the results of the study, when all possible channels of the standard 'observer effect' were excluded [45] and the vision of the near future [46] do exist, but their identification by objective methods is difficult and not always reproducible. This is not surprising since the effects of living consciousness do not have physical causes—they are creative effects, and such effects are always unique and are not reproducible with mechanical regularity. Even in standard scientific studies in biology, pharmacology, and cognitive psychology the repeatability of results is not always guaranteed and decreases over time [47]. Uniqueness and unpredictability generally distinguish the phenomena of living consciousness from physical phenomena. For example, such mundane phenomenon of living consciousness as a night dream is unique and unrepeatable: no dream is repeated twice with the accuracy with which the phenomena of science are reproduced, and yet we see dreams.

Studies of living consciousness show that, in principle, it is possible to go beyond the limits of the Personal Universe in the form of direct communication between living consciousness of different individuals [48] or perception of the immediate future [49, 50]. The effects of living consciousness are weak and do not change the structure of everyday reality in Personal Universe, but their fundamental significance is that they prove the existence of many Personal Universes and the possibility of direct communication between them. Finally, these effects confirm that living consciousness is capable to directly affect objectified consciousness that we call matter. A simplest example is the ability of our thought to act on our brain and our body, and we do not find this empirical fact surprising.

4.8 Ordering the Chaos: Conclusion

The living consciousness of an individual is not derived from or reducible to physical and social nature. The living consciousness is a gift whose origin we do not know. It depends on our internal Self whether this gift will be a pleasant experience for us or turn into hell. With the help of this gift, the person extracts the physical and social universes from nonexistence and forms his or her Personal Universe. But the physical universe does not appear in the form understandable to the person, it appears in the form of Chaos. As the ancient Greek poet said in around 700 BC, "In truth at first Chaos came to be, but next wide-bosomed Earth" [51]. A person must order this Chaos by projecting onto it the properties of his or her living consciousness. In the paradigm of Personal Universe, a person has to make such a projection on his or her own. And if for an infant the primary Chaos is already partially ordered, then a scientist at the forefront of science is dealing with the raw Chaos.

Science is inexorably moving forward. Struck by the perfection of the products of science, some of us begin to ignore the fact that these products could not be made without projecting scientists' living consciousness onto the physical universe. And the origins of living consciousness remain a mystery. Trying to explain the mystery, some scientists resort to the hypothesis that a person could be a character in the cosmic computer simulation. In the next chapter we will review the study on how children and educated adults react to this hypothesis, and argue that, with some adjustments, this hypothesis is not completely ungrounded.

References

1 Peacock, J. A. (1999). *Cosmological Physics*. Cambridge University Press. ISBN 978-0-521-42270-3.
2 The cosmic microwave background radiation. https://aether.lbl.gov/www/science/cmb.html
3 Keel, W. C. (2007). *The road to galaxy formation* (2nd ed.). Springer-Praxis. ISBN 978-3-540-72534-3.
4 Smith, H. A. (2017). Questioning Copernican Mediocrity. *American Scientist, 105*(4), 232. https://www.americanscientist.org/article/questioning-copernican-mediocrity
5 Billings, L. (2020). Do we live in a Lopsided Universe? Scientific American. https://www.scientificamerican.com/article/do-we-live-in-a-lopsided-universe1/
6 Ward, P., & Brownlee, D. (2000). Rare Earth: Why complex life is uncommon in the universe. Copernicus Books. ISBN 0-387-98701-0.
7 https://shop.cosm.org/products/gaia-poster
8 Bostrom, N. (2020). Anthropic Bias. https://anthropic-principle.com/q=book/table_of_contents/
9 Wheeler, J. A. (1990). Information, physics, quantum: The search for links. In W. H. Zurek (Ed.), *Complexity, entropy, and the physics of information* (pp.309–336) Addison-Wesley. https://philpapers.org/archive/WHEIPQ.pdf
10 Faye, J. (2019). *Copenhagen Interpretation of Quantum Mechanics. Stanford Encyclopedia of Philosophy.* https://plato.stanford.edu/entries/qm-copenhagen/
11 Francis, M. (2012). Quantum entanglement shows that reality can't be local. Ars Technica. https://arstechnica.com/science/2012/10/quantum-entanglement-shows-that-reality-cant-be-local/
12 The Nobel Prize in Physics 2022. https://www.nobelprize.org/prizes/physics/2022/summary/

13 Kragh, H. (2013). Big Bang: The etymology of a name. *Astronomy & Geophysics, 54*(2), 28–30. https://doi.org/10.1093/astrogeo/att035
14 Subbotsky, E. (2023). *Examining the psychological foundations of science and morality: Explaining the inexplicable*. Routledge.
15 Blaisdell, A. P., Sawa, K., Leising, K. J., & Waldmann, M. R. (2006). Causal reasoning in rats. *Science, 311*, 1020–1022.
16 Grosenic, L., Clement, T. S., & Fernald, R. D. (2007). Fish can infer social rank by observation alone. *Nature, 445*, 429–432.
17 Langford, D. J., Crager, S. E., Shehzad, Z., Smith, S. B., Sotocinal, S. G., Levenstadt, J. S., Chanda, M. L., Levitin, D. J., & Mogil, J. S. (2006). Social modulation in pain as evidence for empathy in mice. *Science, 312*, 1967–1970.
18 Edgar, J. L., Lowe, J. C., Paul, E. S., & Nicol, C. J. (2011). Avian maternal response to chick distress. *Proceedings of the Royal Society B*. https://royalsocietypublishing.org/doi/https://doi.org/10.1098/rspb.2010.2701
19 Markov, A. (2011). *Evolution of Man*. Volume 2. Apes, Neurones, and the Soul. Corpus.
20 Markov, A. (2015). *The birth of complexity. Evolutionary biology today: Unexpected discoveries and questions*. Corpus (ACT).
21 Piaget, J. (2013). *The construction of reality in the child*. Routledge.
22 Darwin, C. (2019). On the origin of species. The Natural History Museum.
23 https://www.iflscience.com/non-living-things-may-also-evolve-with-their-own-survival-of-the-fittest-71152
24 https://www.greekmyths-greekmythology.com/narcissus-myth-echo/
25 Irwin, W. (Ed.) (2002). The Matrix and Philosophy: Welcome to the Desert of the Real. https://cdn.mashreghnews.ir/old/files/fa/news/1392/9/25/459414_291.pdf
26 Gleiberman, O. (2018). Film Review 'Upgrade'. https://variety.com/2018/film/reviews/upgrade-review-leigh-whannell-1202826627/
27 The Lawnmower Man (film). https://www.rottentomatoes.com/m/lawnmower_man
28 Ramsey, W. (2016). Eliminative materialism. The Stanford Encyclopedia of Philosophy. https://plato.stanford.edu/archives/win2016/entries/materialism-eliminative/
29 Robinson, W. (2013). Epiphenomenalism. The Stanford Encyclopedia of Philosophy. https://plato.stanford.edu/entries/epiphenomenalism/
30 Dennett, D. (2005). *Sweet dreams: Philosophical obstacles to a science of consciousness*. MIT Press.
31 Bostrom, N. (2003). Are you living in a computer simulation? *Philosophical Quarterly, 53*(211), 243–255. https://doi.org/10.1111/1467-9213.00309
32 Kandel, E. R. (2012). *Principles of neural science*. McGraw-Hill Education.

33 Crick, F. (1994). *The astonishing hypothesis: The scientific search for the soul.* Simon & Schuster.
34 Penfield, W. (1975). *The Mystery of the Mind.* Princeton University Press.
35 Rock, A. (2009). *The mind at night.* Basic Books.
36 https://www.explainthatstuff.com/analog-and-digital.html
37 Subbotsky E. (2020). *The Bubble Universe: Psychological perspectives on reality.* Palgrave.
38 Subbotsky, E. (2020). *Faith through the prism of Psychology.* Routledge.
39 Dye, L. (2003). Babies Recognize Mom's Voice from the Womb. https://abcnews.go.com/Technology/story?id=97635&page=1
40 Simion, F., & Giorgio, E. D. (2015). Face perception and processing in early infancy: Inborn predispositions and developmental changes. Frontiers in Psychology. https://www.ncbi.nlm.nih.gov/pmc/articles/PMC4496551/
41 Miller, G. A. (1956). The magical number seven, plus or minus two: Some limits on our capacity for processing information. *Psychological Review, 63*(2), 81–97. CiteSeerX 10.1.1.308.8071. https://doi.org/10.1037/h0043158.
42 Nemeroff, C., & Rozin, P. (2000). The making of the magical mind: The nature and function of sympathetic magical thinking. In K. S. Rosengren, C. N. Johnson, & P. L. Harris (Eds.), *Imagining the impossible: Magical, scientific, and religious thinking in children* (pp. 1–34). Cambridge University Press.
43 Subbotsky, E. (2010). *Magic and the mind: Mechanisms, functions and development of magical thinking and behavior.* Oxford University Press.
44 Groome, D., & Law, R. (2016). Memory Improvement. In D. Groome, M. W. Eysenck (Eds.), *An introduction to applied cognitive psychology* (p. 131). Routledge. ISBN 978-1-317-55652-7.
45 Subbotsky, E. (2013) Sensing the future: Reversed causality or a non-standard observer effect? *SAGE open*, 1–17. https://openpsychologyjournal.com/VOLUME/6/PAGE/81/ABSTRACT
46 Subbotsky, E., & Ryan, A. (2015). Motivation and belief in the paranormal in a remote viewing task. *The Open Behavioral Science Journal, 15.* https://benthamopen.com/ABSTRACT/TOBSJ-8-1
47 Lehrer, J. (2010). December). The truth wears off. Is there something wrong with the scientific method? *New Yorker, 13*, 52–57.
48 Bem, D. J., & Honorton, C. (1994). Does Psi exist? Replicable evidence for an anomalous process of information transfer. *Psychological Bulletin, 115*, 4–18.
49 Roe, C. R., Cooper, C. E., Hickinbotham, L., Hodrien, A., Kirkwood, L., & Martin, H. (2020). Performance at a precognitive remote viewing task, with and without Ganzfeld stimulation: Three experiments. *Journal of Parapsychology, 84*(1), 38–65.

50 Tressoldi, P., & Katz, D. L. (2023). Remote viewing: A 1974–2022 systematic review and meta-analysis. *Journal of Scientific Exploration*. https://doi.org/10.31275/20232931
51 Hesiod, Theogony, line 104. https://www.perseus.tufts.edu/hopper/text?doc=Perseus%3Atext%3A1999.01.0130%3Acard%3D104

CHAPTER 5

Is Living Consciousness a Simulation? Children's and Adults' Reactions to the 'Person Is a Simulation' Hypothesis

5.1 Introduction: Is a Person a Simulation?

The famous 'third law' by the science fiction writer Arthur Clark proclaims 'Any sufficiently advanced technology is indistinguishable from magic' [1, p. 36]. Indeed, many phenomena that in the past were considered magical (e.g., flying in the air and speaking to other people remotely) today became pieces of advanced technology. Many, but not all. The discovery of expansion of the physical universe gave birth to the Big Bang theory, according to which the universe emerged from singularity—the point of infinite dense and gravity—and continues to expand today [2]. The reasons for the 'explosion' are a mystery, as are the emergence of the physical laws [3]. The origins of life in the universe are equally mysterious. In later editions of his book 'The origins of species' Charles Darwin acknowledged that the first forms of life had been 'breathed' into organisms by the Creator. Despite many attempts by evolutionary scientists to explain the origins of life, these origins remain irreducible to known physical laws, and the probability of spontaneous generation of life from unanimous matter is assessed as hopelessly small [4]. The argument that in the infinite time, every probability has a chance to come true contradicts the data of modern cosmology, according to which the universe had a beginning and will come to its end. And there are other mysteries in the universe, such as the 'fine tuning' of the universal cosmological constants and the 'entanglement effect' in the quantum world (see Chapter 4 for

E. Subbotsky, *The Magic of Living Consciousness*,
https://doi.org/10.1007/978-3-031-54708-9_5

more on that). It appears that some mysterious magical force does indeed act in the universe.

How can science resolve this seemingly unescapable influx of magic in the observable world? One of the possible ways to replace magic and God with advanced technology is to present the universe, and a person in the universe, as a result of a digital simulation.

Indeed, in the second half of the twentieth century, at the peak of computer euphoria, there appeared the idea that it is possible to simulate the universe by converting every atom into a series of ones and zeros. Because a person is a part of the universe, every individual and his or her living consciousness would be a part of this cosmic digital program. The American physicist Frank Tipler wrote "There is no way for the people inside this simulated universe to tell that they are merely simulated, that they are only a sequence of numbers being tossed around inside a computer and are not in fact real" [5, p. 35]. How do we know that we are real and not a simulation in some gigantic computer—Tipler asks? And answers: We don't know, but it doesn't matter. Likewise, it doesn't matter whether the universe is real or merely a simulation. All that matters is whether it is possible to create an abstract program that can simulate the whole universe. The corollary of this provocative theory is that there must be a Supreme Programmer out there who can produce computer technology so advanced that it can create a person and make the person believe that he or she is not a computer character but an individual with free will. If Arthur Clark's sentiment is taken seriously, then the Supreme Programmer is indistinguishable from God, and the term 'simulation' in the computer era stands for the traditional term 'creation'. Consequently, in this chapter, the terms simulation and creation will be used as synonyms.

Historically, the idea of a person being a simulation first appears in Plato's dialogue 'Laws', in which one of the characters asks the question: "May we not conceive each of us living beings to be a puppet of the Gods, either their plaything only, or created with a purpose — which of the two we cannot certainly know?" [6]. Plato puts his answer to this question in the same character's mouth: "Man is the puppet of the Gods. He is pulled opposite ways by the strings of his affections but must himself pull against them with the golden cord of reason" [ibid.]. In other words, according to Plato, gods simulate a person's physical body, but not the person's living consciousness, because the person has a free will to decide which of the 'strings' pulled by the gods to follow. In contrast, some modern theorists assume that the whole person, including his or her living consciousness, is a product of some kind of Supreme Programmer—God, aliens, or our distant descendants from the future [7, 8]. However, this assumption raises the following questions: Are the Supreme Programmers a foundational reality, some kind of 'causa finalis', or they themselves are simulated? Do they simulate only our physical body or our internal Self (see [9] for the elaboration on the concept 'internal Self') and living consciousness as well?

At present, the existing AI can simulate the image of a person's body and his or her 'hard' consciousness—logical operations—but not the person's living consciousness. The characters of cartoons, movies, and computer games are puppets controlled by the programmer and don't have internal Selves of their own (since the internal Self is a holder of living consciousness, in this chapter the term 'internal Self' will stand for both internal Self and the living consciousness). Even the most powerful neuro-nets, such as OpenAI, which can generate realistically looking visual animations just from verbal description, make only physical representations of reality, which is devoid of the psychic dimension. So, if we accept that the 'person is a simulation' hypothesis is confined to physical domain (i.e., human bodies), then the question arises where our internal Self comes from. Indeed, when I say "I am thinking therefore I exist"—what do I mean by the 'I'—a created thing, a thing that evolved, or something other?

The religious answer is that God created a person's body and breathed the living consciousness into it. In contrast, evolutionary scientists insist that living consciousness evolved in the course of evolution [10]. Clearly, in the construction of these answers our internal Self is involved, which makes these answers circular. The attempt to get out of the circularity is to make the internal Self a simulator of the external world, which most clearly was done by Immanuel Kant. According to Kant, we perceive the world by turning 'things in themselves' into phenomena based on the innate properties of our living consciousness, such as time, space, and causality [11]. But putting our internal Self 'first in the line of creation' doesn't answer the question of its origins.

Indeed, as argued in Chapter 4, our internal Self is an entity unobservable to itself. In other words, we don't have an external perspective from which we could observe our internal Self to convert it into a scientific concept as we do with physical objects. Earlier we suggested that our internal Self belongs to the domain of the Unknowable (see [9,

Chapter 5] for the elaboration on that). Still, we constantly use the term 'I' for reasoning about our free actions and know that our internal Self has a finite existence: it appeared at some point and will inevitably disappear. It also disappears in sleep without dreams and in the state of general anaesthetics. Note that in such states our living consciousness is still active on the subconscious level, and this differs the states of unconsciousness from death. The fact that our internal Self is impermanent (it appears, disappears, and reappears) means that there must be an entity, which is a Holder of our internal Self, and this Holder must necessarily be a permanent entity, of which our internal Self is only a temporary and imperfect reflection [ibid.]. Both the conceptual and empirical analyses show that our body cannot be such a holder, since for the unconscious person his or her body doesn't exist, and the indirect testing of whether the living consciousness of the person under general anaesthetics is still functioning (e.g., by asking the patient questions to which the patient can 'answer' by squeezing the doctor's hand) revealed that the patient's living consciousness could still be present even when all the physiological indicators of the presence of living consciousness are absent (see [9], Chapter 4, Section 4.2). This leaves one with the option, according to which the Holder must be a permanent source of our living consciousness, to which our brain/body system is permanently connected while we are alive, but can partially relax this connection to switch our internal Self on and off, depending on whether we are conscious or not.

In the everyday life, we call this Holder God. A usual objection to acknowledging the existence of God is that the idea of God is purely imaginary, like the idea of a dragon or a unicorn, and from having the idea of God it doesn't follow that this idea has a referent in the real world. The counterargument to this objection is that we imagine God as an entity with attributes such as omnipotence, and this implies the capacity of the omnipotent entity to obtain reality in the world independently of our imagination, whereas all other imaginary entities have a limited power and do not necessarily exist outside our mind. So, in our Personal Universe the conclusion that God we are imagining as an almighty entity must exist independently of our imagination necessarily follows from the existence of our living consciousness. Switching from religious to scientistic terminology, one can say that from the ability of our living consciousness to produce the idea of the almighty Supreme Programmer necessarily follows the conclusion that such Supreme Programmer must exist outside our living consciousness. The logical necessity of the independent existence of the Supreme Programmer is supported by the empirical fact that the existence of our living consciousness is periodically interrupted by the states of unconsciousness. In other words, the sheer fact that we can have the idea of the almighty Supreme Programmer capable of creating our internal Self makes the real existence of such Supreme Programmer necessary (see [12] for more on that). And vice versa, the acknowledgement of the independent existence of the Supreme Programmer capable of creating our internal Self implies that our internal Self could indeed be created.

Accordingly, the study reviewed in this chapter aimed to examine whether children and educated adults could infer that the Supreme Programmer must exist independently of their imagination in external reality from the fact that they have the idea of the almighty Supreme Programmer in their living consciousness.

The main problem in examining this was to find a suitable method. The first difficulty in developing such a method was finding the name for the Supreme Programmer. As stated earlier, in popular culture the name for the Supreme Programmer is God. According to the Bible, creation of the person was a simulation of God's imagination (e.g., And God said, "Let us make man in our image, after our likeness": Genesis, 1:26). In the light of this one might suggest to employ the term 'God' in dialogues with children.

The negative response to this suggestion is rooted in a specific cultural load that the term 'God' has for a child educated in the European cultural context. If directly asked about the existence of God, the children would be encouraged to reveal their socially accepted beliefs about God rather than their independent thinking of the existence of the Supreme Programmer. Thus, the children brought up in a Christian culture would be tempted to say that God exists in reality, whereas children educated in the atheistic tradition, like the one in the former Soviet Union, would be likely to deny the God's independent existence [13]. Hence, children's answers that reflect their adherence to prevailing views could easily be mistaken for their independent views on the existence (nonexistence) of the Supreme Programmer beyond the domain of their imagination.

This was the main reason why the term 'Almighty Wizard', rather than God, was taken as a substitute for the Supreme Programmer. Other reasons included the fact that this term was in active vocabulary of both children and adults and that this term, by and large, contained the required property of almightiness: the ability to obtain existence independent of our living consciousness.

Another difficulty with eliciting the children's authentic views on whether they are or aren't created by the Almighty Wizard was the way of putting the problem. Obviously, asking participants directly the question of whether they were created by the Almighty Wizard could reveal their culturally embedded image of the term 'wizard' rather than their independent thinking about the problem. Both in the Russian and English cultures a wizard is a fictitious character: wizards operate in fairy tales in which they perform magical things, but they don't create children in the real world.

More suitable would be to face the participants with the problem indirectly. To accomplish this, the problem was put as a question about the ability of the Almighty Wizard that participants were imagining to leave their imagination and come out into the real physical world. If a participant accepts the suggestion that the Almighty Wizard who can create a person like him or her could leave his or her imagination and obtain independent existence, then the participant indirectly acknowledges that he or she may have been created by the Almighty Wizard.

5.2 How Mighty Is the Almighty Wizard? Children's Reaction to the Idea That the Almighty Wizard Could Come Out of Their Imagination into the External World

The study was in the form of an interview [14]. Russian children in Moscow in the middle of 1980s and British children in Lancaster, United Kingdom, in the early 1990s, aged 4, 5, 6, 7, 9, and 11 years, were invited to the interview.

The interview included three stages. The *first stage* examined whether the children were ready to acknowledge the existence of the Almighty Wizard in the external physical world and not only in the domain of the imagination (fairy tale and imaginary play). Among other questions, the following questions were asked:

1. "Is there anywhere in the real world a person or a fairy tale wizard who knows everything and can do everything he wants? Can such Almighty Wizard exist in a fairy tale? Can he exist within a play?"

The results indicated that almost all children, except for a few 4-year-olds, denied that the Almighty Wizard could exist in the real world. At the same time, all children acknowledged that such a character could exist in their imagination as a character of a fairy tale or a pretend play.

The aim of the *second stage* (questions 2–3) was to determine the limits of the Almighty Wizards' power, specifically, whether the children thought that almightiness included the capacity for the wizard to move from the domain of the imagination into the perceived world:

2. "Is this Almighty Wizard capable of creating you or somebody like you?"
3. "Could this Almighty Wizard come out from a fairy tale or our imagination into the real world?"

The second stage of the interview revealed that all children agreed that 'inside a fairy tale' the wizard could create everything, including a living child, but almost all of them denied that the imaginary wizard could cross the fairy tale boundaries and enter the external world.

In the *third stage* of the interview, the children were put through a series of steps that aimed at making them aware of the link between the imaginary wizard's almightiness and his ability to come out of the imagination and obtain independent, and not just imaginary, existence:

4. "Can you imagine such an Almighty Wizard who would be able to come out from your imagination in the real world?"
5. "Please, imagine such a wizard right now, O.K.?"
6. "Now, if you have the idea of the Almighty Wizard in your mind, does this wizard exist in your mind?"
7. "But if this wizard is almighty and he exists in your mind, could he come out of your mind in the real world, for instance, in this room and sit down over there on the chair?"

The results showed that most children agreed to imagine a wizard that could go out of their living consciousness into the real world. Nevertheless, the children unanimously denied that such an Almighty Wizard could really jump out of their minds and sit down in the chair in front of them.

To examine whether external assistance in the form of logical priming could make the children realize that imagining the Almighty Wizard includes the Wizard's ability to acquire independent existence, the logical contradiction in the children's judgements between the wizard's almightiness and his simultaneous disability to do something was exposed to them via discussion. The children were repeatedly asked to imagine wizard number two (three, four, five) who is even more powerful than the wizard of the previous number and who "really could jump from their imagination into this room". However, even when the wizard's might had been increased four times, this failed to shatter the children's certainty that the Almighty Wizard is unable to leave their imagination and obtain existence independent of their living consciousness.

For example, Natasha (a girl, 6 years) answered the question about the wizard's ability to leave her mind for the perceived world as follows "No, because he can only exist in our thoughts or in books, but he can't go out into the real world".—Exp. "Therefore, he is not almighty, is he?"—"No, even if they write in books that he can do everything, it is not true, it is just for fun".—Exp. "O.K., but can we think of such an almighty wizard who can really come out of our imagination in the real world, or we cannot?"—"Yes, we can think, but he won't be able to go out of our thoughts".—Exp. "Therefore, he isn't almighty?"—"They write in books that he is almighty, but in reality he isn't, and he doesn't exist".

Another girl, Nastija (7 years) agrees that if the wizard is 'really almighty' he must be able to really go out of our head and sit down on the chair. However, when the experimenter asked "Therefore, if we strongly think of him coming out from our head, will he come out and sit down here in front of us?" the girl changed her mind, saying "No, because he can't go out, he has no power enough to go out of our thoughts". A similar answer is given by Oleg (boy, 7 years) "No. We are just thinking of him and imagining that such a person exists, just like those people who believe that there is God in the world, but in reality, there is no God".

Replication of this study with British children in the United Kingdom didn't reveal any significant differences between the answers of Russian and British children. Altogether, the results showed that most children of all age groups in Russia and the United Kingdom strongly resisted the idea that the Almighty Wizard, who can create a person like themselves, could obtain independent existence. The question arises of what caused the resistance?

A possible cause was that the children were unable to make the inference that the attribute of almightiness implies the ability of the almighty character to obtain existence independent of their living consciousness.

However, there could be alternative explanations. To examine these alternative explanations, a series of interviews was conducted with graduates and postgraduates of Lancaster University, United Kingdom.

5.3 Jumping Out of the Imagination: Adults' Responses to the Suggestion that the Almighty Wizard Could Obtain Independent Existence

The aim of the interviews was to explore the extent to which adults can accept that the Supreme Programmer is not a product of pure imagination but must exist outside their living consciousness. It was assumed that recognizing the reality of the Supreme Programmer is tantamount to recognizing that a person could be the product of the Supreme Programmer, otherwise we wouldn't have empirical evidence proving that the Supreme Programmer could create a person.

Like in the interviews with children, in the interviews with adults the term for the Supreme Programmer was the 'Almighty Wizard'. The retention of the term Almighty Wizard made it possible to compare the results of the interview with adults with those conducted with children, and at the same time eliminated the danger of direct identification of the Supreme Programmer with God.

In Interview 1, the questions were the same as in the interview with the children, with the key question being whether the Almighty Wizard could or could not leave their imagination and acquire independent existence. If the key question was answered in the negative, the experimenter asked two additional questions "Do you agree or disagree that there is a logical contradiction in what you have said?" and if the answer was "yes", then the key question was repeated "Do you prefer to accept this contradiction, or would you like to change your mind and agree that the Almighty Wizard is able to step out of your imagination and enter this room?" The purpose of the additional questions was to assist the participants in becoming aware of the fact that the attribution of almightiness to the imaginary Wizard logically includes the Almighty Wizard's ability to obtain independent existence.

If after the additional questions, the participants did not recognize the logical contradiction in their judgements, then they were classified as insensitive to external assistance. However, if the participants admitted that their denial of the Almighty Wizard's ability to step out of their imagination into the real world put them in a logical contradiction with themselves, they were still left with two options to answer the key question.

On one hand, they could change their minds and accept the conclusion that the Almighty Wizard could indeed step out of their mindset into the perceived world; if the participants chose this answer, they were effectively admitting that the 'person is a simulation' hypothesis might in fact come true. On the other hand, participants could accept the contradiction while continuing to deny the ability of the Almighty Wizard to get out of their living consciousness into the external world. Participants who choose the second answer 'de facto' deny that the 'person is a simulation' hypothesis can be true.

The results of Interview 1 showed that all participants admitted that they could imagine the Almighty Wizard who is able to leave their imagination and enter the perceived world, however, refused to accept the suggestion that the Wizard could indeed leave their imagination and obtain independent existence.

After additional questions, one of the seventeen participants admitted the fact of the real existence of the Almighty Wizard ("He is here, but we cannot see him. He does not emit light waves"), but the rest categorically denied this. Of the remaining sixteen participants, eleven did not see a logical contradiction in their judgements, and five said that there was a contradiction, but chose to accept it and continued to deny that the Almighty Wizard could come out of their imagination into the perceived world.

Altogether, the results showed that educated adults, like the children, denied the ability of the Almighty Wizard to go beyond their imagination. The vast majority either did not recognize the logical contradiction in their judgments about the impossibility of the Almighty Wizard to enter the external world or preferred to put up with this contradiction rather than release the Almighty Wizard into the external world.

From these results, it follows that the obstacles to recognizing the independent, and not just imagined, existence of the Almighty Wizard were psychological rather than based on logical inferences. To rule out the most likely of these obstacles, three more interviews were conducted, each with a different group of participants. In each of these interviews, the main section, which was by and large identical to Interview 1, was preceded by a prompting section aimed at addressing a possible reason for the participants' reluctance to release the Almighty Wizard into the external world.

One of the possible psychological barriers to accepting the independent reality of the Almighty Wizard was that the participants viewed the Almighty Wizard's ability to come out of their mind as a perverse causal event. Since the Almighty Wizard had unlimited creative power, the participants might have thought that they were being pushed into admitting a perverse causality sequence: The entity with unlimited creative power is created by a person with limited creative power.

To prevent this notion from arising, in the prompting section of Interview 2 the participants were helped to realize that it could in fact be the Almighty Wizard who had created the participants and then simply 'settled down' in their minds in the form of their thinking (speaking) about him. The realization of this was supposed to make clear to the participants that it was not their imagination that created the Almighty Wizard, but that the Almighty Wizard used his unlimited capacity to do things to implant the idea of the Almighty Wizard in their imagination.

The questions in the prompting section of Interview 2 were as follows:

1. Now I would like to talk to you about wizards, okay? Have you ever heard or read anything about them? What have you heard (read)?
2. Tell me, if the wizard is almighty, could he create a human person, yes, or no?
3. Could this Almighty Wizard create a person like yourself?
4. And could this Almighty Wizard create some thoughts in your head, for example, could he make you think about the Almighty Wizard, yes, or no?
5. What do you think, could this Almighty Wizard, if he is almighty, turn himself into your thoughts, enter your mind and settle down there, yes, or no?
6. If this happened, would you know that this Almighty Wizard settled down in your mind, or would you think that it was yourself who gave rise to thoughts about this Almighty Wizard?

If in Interview 1 the psychological obstacle that prevented the participants from releasing the Almighty Wizard from their imagination into the outside world was the participants' resistance to accept the perverse causal sequence, then in the main section of Interview 2 the participants would be more willing to acknowledge the ability of the Almighty Wizard to get out of their living consciousness into the external world than in Interview 1.

Another psychological obstacle to accepting independent reality of the Almighty Wizard could be anthropomorphising the image of the Almighty Wizard. It is possible that the participants viewed the Almighty Wizard as a humanoid being with a physical body. In such a case, it would seem obvious to them that, no matter how hard they tried, it was very unlikely that such a humanoid figure would appear in front of them in the flesh.

To prevent the possibility of anthropomorphizing the Almighty Wizard, in the prompting section of Interview 3 participants were led to the realization that the Almighty Wizard could be invisible and not necessarily human.

The following questions were asked:

1. Now I would like to talk to you about wizards, do you agree? Have you ever heard or read anything about them? What have you heard (read)?
2. Tell me, is it mandatory that wizards look like people, or can they look differently?
3. Could a wizard, if he is almighty, be invisible, yes, or no?
4. Could the wizard turn himself into various animals and objects, yes, or no?
5. Could the wizard turn himself into air, or into the walls of this room, yes, or no?
6. So, if the wizard is almighty, he is not necessarily visible and could be hidden in various objects in this room, yes, or no?

If the participants' reason for their refusal to release the Almighty Wizard from their living consciousness into the external world was the anthropomorphizing the Wizard's image, then in the main section of Interview 3 the number of participants who acknowledged the ability of the Almighty Wizard to get out of their imagination into the external world would be significantly greater than in Interview 1.

One more psychological obstacle could be the participants' belief in the 'internal/external barrier'—the impossibility of the imaginary entity to instantly obtain independent existence in the external world. The belief in the internal/external barrier means that to transform a mental image into a perceived physical object certain conditions must be met, such as the presence of some kind of external 'primary matter', tools, and special efforts applied to the primary matter. For example, if a sculptor decides to make a statue out of a piece of stone, he or she needs the stone, a chisel, and a hammer, and then must work on the piece of stone with these tools. As a result, the statue, that exists in the sculpture's living consciousness as a mental image, acquires its external embodiment—a part of the objectified consciousness.

Violation of the internal/external barrier would mean that instead of a causal event (the living consciousness acting on a person's body to create a physical object with the help of physical tools) a mind-over-matter magical event occurred: A person's mental image instantly and without any physical efforts of the person obtains an external embodiment. In the earlier work we called this type of event participatory causality: Although in regard to our body participatory causality is our everyday experience exhibited in every voluntary action, the prevailing view in the modern industrial world denies that participatory causality could be at work in regard to external physical objects (see [9], Chapter 7 for the elaboration on the difference between physical and participatory kinds of causality).

Since the idea of the Almighty Wizard is a part of the participant's living consciousness, they could treat this idea like any other imagined object (for example, an imagined house) and apply the belief in the internal/external barrier to it. Just like a desire to have a nice house cannot result in the immediate physical emergence of the house, the idea of the Almighty Wizard cannot result in the emergence of the Almighty Wizard in the external world.

To suspend the belief in the internal/external barrier, in the prompting section of Interview 4 participants were individually shown a phenomenon in which an object that existed in the participant's imagination instantly 'transformed' into a perceived physical object. The aim of the demonstration of this mind-over-matter magical phenomenon was to suspend the participants' belief that an instant conversion of an imagined object into its external embodiment is impossible.

The procedure for the prompting section of Interview 4 was as follows. The participant was given an empty wooden box in his or her hands, offered to carefully examine it and asked to close the lid of the box. The experimenter then showed the participant a postage stamp with a certain image on it, which was immediately removed from view, and asked the participant to make a wish for this postage stamp to appear in the box.

Upon opening the box, which was still in the participant's hands, the participant discovered that the postage stamp he or she wanted to emerge in the box was indeed in the box. The participant was asked to examine the box once again for the presence of a hidden mechanism in it. Having found nothing suspicious, the participant entered the state in which his or her belief in the internal/external barrier was suspended, after which the participant was asked questions from the main section of the interview.

If the psychological factor that prevented the participants from recognizing the possibility for the Almighty Wizard to exit their living consciousness and enter the external world was their strong belief in the internal/external barrier, then suspending this belief would increase the number of participants recognizing the Almighty Wizard's ability to get from their imagination and obtain independent existence in the main section of Interview 4 compared with Interview 1.

The results showed that the number of participants who admitted the ability of the Almighty Wizard to get out of their living consciousness into the external world in Interview 2 and 3 was not statistically different from that number in Interview 1, but in Interview 4 it was (Table 5.1).

Table 5.1 Number of adult participants who denied or accepted the idea that the Almighty Wizard might come from their imagination into the real world, as a function of *Interview*

	Interview			
Number of participants	*1*	*2*	*3*	*4*
Denied	16	14	15	7
Accepted	1	1	3	5
Differences between numbers of adult participants who accepted the idea that the Almighty Wizard might come out of their imagination into the external world				
Between interviews		1–2	1–3	1–4
Chi-Square value		0.008	1.005	5.489
Significance		n/s	n/s	$p < 0.05$

5.4 Why Is It Hard to Accept that a Person Is a Simulation: Conclusion

The results of the experiments showed that under normal circumstances both children and adults strongly opposed the possibility that the almighty Supreme Programmer who is able to simulate them could obtain independent existence. The fact that almost all participants either denied the logical contradiction in their judgement or preferred to accept the logical contradiction rather than allow the Supreme Programmer to exit their imagination and enter the external world showed that the reasons for the denial were psychological, rather than logical. The investigative experiments with adults showed that psychological factors such as the perverse causality and the anthropomorphizing the image of the Supreme Programmer were not the reasons for the denial, while the belief in the internal/external barrier was. Indeed, out of the three aforementioned psychological factors only suspending the participant's belief in the impossibility of a direct conversion of a mental image into its physical 'doppelganger' made a significant number of participants recognize that the Supreme Programmer (coded in the interviews as Almighty Wizard) could get out of their imagination into the external world. Such recognition is a tacit acknowledgement of the possibility that the 'person is a simulation' hypothesis might be true.

Along with providing new empirical data on the popular reaction towards the 'person is a simulation' hypothesis the results of this experiment highlight the implication of this hypothesis that thus far has been mostly ignored. More specifically, the scientists' attempt to escape the mystical element in interpreting the origins of living consciousness via suggesting that a person is a simulation is based on the tacit admittance that *overcoming the internal/external barrier through the magical participation must be possible in the real world*. Indeed, to be able to simulate a real person from a mental image of this person the Supreme Programmer must have the ability to convert its mental images into external reality. At that, there should be a difference between simulating the person's body and simulating the person's living consciousness. Since a person's body is a part of the physical universe and consists of physical elements, it could in principle be digitalized. In fact, an artificial limb supplied with the elements of AI could be interpreted as a simulation, albeit rather imperfect, of a real limb, with the human programmer being a cause, and the simulated artificial limb being an effect. In principle, it is possible to imagine a Supreme Programmer, which could create a program capable of building a human body from available physical matter. In fact, the mysterious appearance of the two rRNAs with a total size of at least 1000 nucleotides required for the launch of a living cell [4] looks very much like a kind of simulation by some super-intelligent being on the 'cosmic computer', and the rest could be done by evolution through natural selection.

However, to simulate the human living consciousness the Supreme Programmer doesn't have the 'raw material' except the living consciousness of his own, and the living consciousness is a mental reality inaccessible to digitalization (see Chapter 4 for more on that); consequently, to simulate living consciousness of a person, the Supreme Programmer must share part of his or her own living consciousness with the human body that he or she had created. As the Bible shrewdly says, "And God formed dust from the earth into a man and he blew into his face a breath of life, and the man became a living soul" [Genesis 2:7]. As a result, the simulation of the living consciousness of a person cannot be a causal event; instead, such a simulation must be establishing the relation of a magical participation between the living consciousness of the Supreme Programmer and the living consciousness of the person (see [9], Chapter 5 for more on that).

That is the reason why existing achievements of digital simulation are limited to simulation of physical objects and routine mental operations of the 'hard' objectified consciousness (see Chapter 1 for more on the division between the 'hard' and 'soft' consciousness), but even the most advanced computers are unable to simulate subjective experiences of living consciousness, such as feelings or unprogrammed goal setting. For simulation of the living consciousness, the simulator must be able to 'implant' part of his or her living consciousness into the 'digital body' of a simulated character (e.g., of Alice in the Wonderland) via magical participation. The results of this experiment revealed that accepting the idea of such direct 'implantation' of living consciousness into external reality is considered by most people as impossible.

To reiterate, the study has shown that the main psychological factor that prevented participants from recognizing the possibility for the Almighty Wizard to move out of their living consciousness into the external world was the belief in the internal/external barrier—the impossibility of the magical participation between the participants' living consciousness and external reality. This belief was strong enough to overcome the participant's belief in the laws of formal logic.

One might assume that the internal/external barrier is relevant only to the barrier between mental and physical realities, and that participants in this study simply objected to the possibility of transformation of their mental image of the Almighty Wizard into a physical body. However, Interview 3 with adult participants ruled this assumption out: this interview showed that even when participants had acknowledged that the Almighty Wizard could be a non-physical invisible entity hiding itself in physical objects available in the experimental room most participants kept insisting that the Almighty Wizard couldn't leave their imagination. It was only when the participants' disbelief in that their living consciousness could magically create/simulate an object in the external physical world was suspended in Interview 4, thus demonstrating that the magical participation between a person's living consciousness and external reality was in fact possible, that a significant number of participants acknowledged the Almighty Wizard's ability to obtain independent existence.

To summarize, the reviewed study revealed that the disbelief of a modern rational person in that magical participatory-correlational causality could work in his or her Personal Universe is a reason of why the 'person is a simulation' idea, which is the foundation of most religions, meets a strong resistance when presented to people in the context of logical reasoning. Rather than acknowledging the independent reality of the Supreme Programmer, with whom a person's living consciousness is in the participatory-correlational link, most people view such Supreme Programmer as an imaginary creation locked within their living consciousness. As a result, in the modern world not only atheists, but theists as well, consider the 'person is a simulation' hypothesis as a theoretical possibility, and the popular expression "our Creator is always with us" is taken as a metaphor rather than an empirical fact. It is only due to another property of people's living consciousness—faith—that religious beliefs still hold the ground in modern rational societies (see [12] for more on that).

The study presented in this chapter was based on the assumption that in reality the Creator is here with a person at every moment of the person's conscious life, but he (or she) decided to make his or her presence inaccessible to our understanding, by putting a psychological barrier that prevents us to see the presence. Why the Creator did so we can only guess, perhaps, in order to make a person rely on his or her own mind instead of constantly begging the Creator for help and guidance.

Only in rare situations, when listening to spiritual music or observing the glory of the universe, can we intuitively sense the presence of the Creator. It is not in our power to completely eliminate this psychological barrier, but it is possible to detect the presence of this barrier by using the ability of our living consciousness to creative imagination. In reality, the fact that participatory-correlational causality is present in the everyday life is evident from the empirical fact of the brain–computer interface (BCI), which is a direct communication pathway between the mind and an external device through the brain's electrical activity, most commonly a wheelchair or a robotic limb. The participatory-correlational causality is revealed by any voluntary action of our body, which is one of the 'wonders of the mundane'. In the times far away from the modern technical wonders scottish philosopher David Hume (1711–1776) wrote "...is there any principle in all nature more mysterious than the union of soul with body; by which a supposed spiritual substance acquires such an influence over a material one, that the most refined thought is able to actuate the grossest matter? Were we empowered, by a secret wish, to remove mountains, or control the planets in their orbit; this extensive authority would not be more extraordinary, nor more beyond our comprehension" [15, E 7.11, SBN 65]. It is in the form of the voluntary action that our living consciousness directly affects physical reality and obtains external embodiment. However, in the present time the voluntary action is devaluated by the reductionist theories of neuroscience (see Chapters 4 for more on this) and not viewed by most modern rational individuals as the proof in favour of the divine origins of their living consciousness. This makes it necessary in the next chapter to consider the historical development of people's views on living consciousness in more detail.

References

1 Clark, A. (1962). *Profiles of the future: An enquiry into the limits of the possible*. Indigo.
2 Alpher, R. A., & Herman, R. (1988). Reflections on early work on 'Big Bang' cosmology. *Physics Today, 41*(8), 24–34.
3 Barkana, R., & Loeb, A. (2001). In the beginning: The first sources of light and the reionization of the universe. *Physics Reports, 349*(2), 125–238. arXiv:astro-ph/0010468.

4 https://elementy.ru/nauchno-populyarnaya_biblioteka/434611/Veroya tnost_zarozhdeniya_zhizni
5 Tipler, F. J. (1989). *The Omega point as Eschaton: Answers to Pannenberg's questions for scientists.* http://www.geocities.ws/theophysics/pdf/tip ler-omega-point-as-eschaton.pdf
6 The dialogues of Plato, in 5 volumes. Vol 5 "Laws". https://oll.libertyfund.org/title/plato-dialogues-vol-5-laws-index-to-the-writings-of-plato
7 Hoffman, D. S. (2020). *The case against reality: How evolution hid the truth from our eyes.* Penguin.
8 Bostrom, N. (2003). Are you living in a computer simulation? *Philosophical Quarterly, 53*(211), 243–255. https://doi.org/10.1111/1467-9213.00309
9 Subbotsky, E. (2023). *Examining the psychological foundations of science and morality: Explaining the inexplicable.* Routledge.
10 Ginsburg, S., & Jablonka, E. (2022). *Picturing the mind: Consciousness through the lens of evolution.* MIT Press.
11 Kant, I. (2008). *Critique of pure reason.* Penguin Classics.
12 Subbotsky, E. (2022). *Faith through the prism of psychology: A new framework for existentialism.* Routledge.
13 Corriveau, K. H., Chen, E. E., Harris, P. L. (2014). Judgments about fact and fiction by children from religious and nonreligious backgrounds. *Cognitive Science,* 1–30.
14 Subbotsky, E. (1996). *The child as a Cartesian thinker: Children's reasonings about Metaphysical aspects of reality.* Psychology Press.
15 https://davidhume.org/texts/e/7

CHAPTER 6

Living Consciousness in Historical Context

6.1 Beliefs About Living Consciousness in the Era of Modernity and Their Bifurcations in the Era of Postmodernism

To understand reasons why living consciousness today is viewed by some as a mystery, and by others as a shadow of the brain processes, we need to look back in history [1].

According to the prevailing views, until the twentieth century, European culture lived on the ideas of living consciousness laid down by the era of modernity—the historic period that arose in the age of Renaissance in the sixteenth century, continued through the age of Enlightenment, and ended by the late twentieth century. Among others, the basic beliefs about living consciousness in the time of modernity included the following ones: (1) living consciousness reflects reality that is independent of living consciousness (*Realism*); (2) living consciousness is transformed into objectified consciousness via symbols (*Symbolism*); and (3) living consciousness has a measurable degree of free will (*Limited freedom*) [2].

Today we live in the era of postmodernism [3]. There are reasons to assume that approximately in the middle of the twentieth century, the process of bifurcation began, when postmodern alternatives sprouted from the basic beliefs, and the basic beliefs were strengthened and sharpened (Fig. 6.1). Let's consider each of these bifurcations in more detail.

E. Subbotsky, *The Magic of Living Consciousness*,
https://doi.org/10.1007/978-3-031-54708-9_6

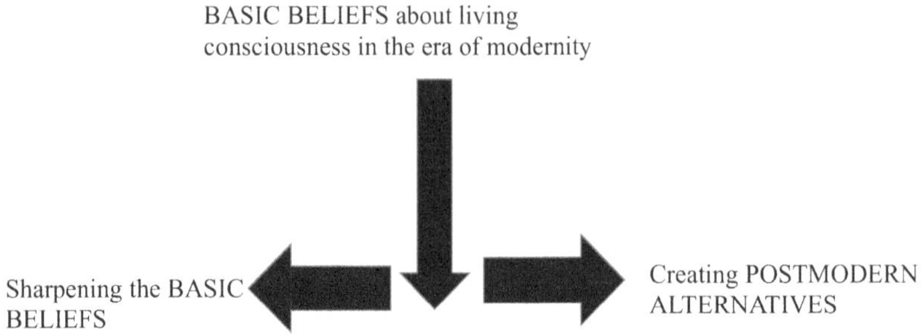

Fig. 6.1 Bifurcations of the BASIC BELIEFS about living consciousness of the era of modernity

6.2 Bifurcation 1: Realism

6.2.1 Basics of the Belief in Realism

Since ancient times, people believed that there are things out there which they experienced with their sense organs. German philosopher Immanuel Kant called such experiences phenomena. The phenomena have colours and shapes, they smell and taste, they are heavy or light, big or small, loud or silent, smooth or rough. They can be chaotic like a sound of traffic, or beautifully organized like a Mozart's symphony.

To meet the definition of a *phenomenon*, subjective experience of our living consciousness must have four basic properties, with the first of them being *identity*. Identity means that the subjective experience has some features that make the phenomenon a unique entity distinct from other experiences. Like other fundamental criteria, identity can only be registered in degrees. Thus, it is impossible to find two identical persons, but two atoms of hydrogen are almost identical, yet their location in space/time still makes them different from each other. Linked to identity is the property of *permanence*, which requires that the subjective experience maintains its identity throughout a certain period and cannot instantly change into another subjective experience. The third property is *locality*: the subjective experience needs to be able to resist merging with other phenomena. According to this property, a phenomenon (e.g., a bullet) cannot permeate another phenomenon (e.g., a human body) without causing destruction in it. Finally, the last property is *inclusion*. According to this property, each phenomenon emerges from preceding phenomena and, having lived for a certain amount of time, turns into some other phenomena.

From the beginning people realized that phenomena are unstable and could be misleading. In order to tame the unstable phenomena, they employ the procedure of comparison, by applying a certain object about which they agreed to use it as a measure, for example, a bar of wood of a certain size. By putting this bar to the building in a standard way, they count the number of times this bar fits to the foundation and the height of the building and then designate the buildings sizes as, say, 20 by 50 bars. By doing this, they created the simplest symbolic substitute, or a *rational construction* (from Latin ratio—reason, understanding) of the building. Rational constructions (RCs for short) include scientific concepts, schemes, blueprints, and other symbolic representations of phenomena. RCs make the ever-changing world of perceived things stable: Whatever distance a person has between herself and the building, and however different the other person's angle of view at the building might be, the building's rational construction remains the same—20 × 50.

Modern science emerges as a discipline that specializes in generating RCs for nearly every object and process of the external and internal worlds: galaxies and black holes, stars and planets, mountains and rivers, plants and animals, our bodies, perceptual organs and brains. Science created RCs not only for objects that we perceive, but also for subjective properties of the living consciousness: subjective space found its objectified referent in physical space, subjective time was converted into physical time, subjective causality solidified into physical causality, and creative thinking cooled down into logical thinking.

As a result, our world acquired a dual structure: The world of our 'soft' living consciousness (phenomena) and the world of the objectified 'hard' consciousness (RC's) (see Chapter 1 for more on the distinction between living and objectified consciousness). The difference between these worlds is that the former we can feel (e.g., we can see, smell, and eat the apple), whereas the latter we can understand (e.g., we can see and understand the word 'apple' and the picture of an apple, but we can't smell and eat it).

As stated above, one of the basic ideas about living consciousness in the era of modernity was the belief in realism. This belief meant that phenomena, with some imperfections, reflect reality independent of the mind. From this it followed that RCs reflect reality too, and because RCs are built by reducing the phenomena's imperfections, they reflect reality even better.

6.2.2 *Bifurcation of the Belief in Realism*

However, beginning with the works of Rene Descartes (1596–1650), followed by works of John Locke (1632–1704), George Berkeley (1685–1753), and David Hume (1711–1776), the idea was taking momentum that phenomena, rather than reflect external reality, reflect properties of our living consciousness. This tendency of *epistemological scepticism* culminated in the teaching of Immanuel Kant (1724–1804), who identified true reality with *noumena*—imagined entities that are inaccessible to our senses. By this division, Kant erected an impermeable watershed between phenomena and reality. As long as phenomena exist in perception, and perception is based on innate properties of our living consciousness, such as subjective space, time, and causality, phenomena too became to be viewed as products of the living consciousness.

There appeared a problem: If phenomena do not reflect independent reality, then RCs, which are built on the base of phenomena, don't reflect reality either. As long as neither phenomena nor RCs reflect independent reality, the troubling question arose of *what is reality*?

There were two ways out of this epistemological puzzle. One way was to give the RCs *ontological primacy over phenomena* and declare RCs to be the only true reality, thus viewing phenomena as a product of RCs (*epistemological reductionism*). Clearly, epistemological reductionism brings us into a logical circle: phenomena, manipulation with which is the way of creating RCs, become reflections of their own products. The alternative way was to *give ontological primacy to phenomena* and view phenomena of living consciousness as the only true reality (*epistemological scepticism*). According to this view, living consciousness does not reflect 'reality as such', but *generates* a 'picture' of reality according to the laws of living consciousness (Fig. 6.2). The belief in epistemological scepticism doesn't mean that our internal Self controls all of the picture of reality it generated. In fact, most of external reality, such as our perceptions of physical and social universes, and part of internal reality, such as our feelings and dreams, are not under control of our internal Self, yet all of reality are brought into *actual existence* by our living consciousness. Anything that is not experienced by our living consciousness, in the form

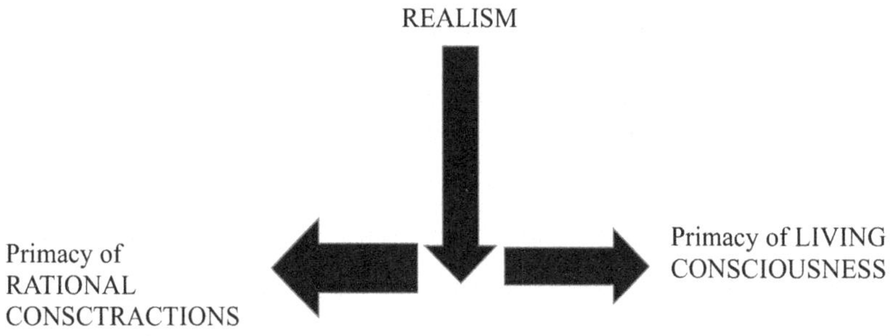

Fig. 6.2 Bifurcation 1: Realism. The basic belief in REALISM bifurcated into the beliefs in primacy of LIVING CONSCIOUSNESS (epistemological skepticism) and primacy of RATIONAL CONSTRUCTIONS (epistemological reductionism)

of perception, thinking, or imagination, belongs to Nothing—the incomprehensible domain of potentialities that exists in our 'hard' consciousness as an empty concept.

Till the twentieth century, the ideas of epistemological scepticism have been circling in a thin layer of intellectuals and did not have a noticeable impact on wider culture. Both scientists and artists of modernity strongly believed in realism. In the twentieth century the situation changed, with art and science opting for different ways.

6.2.3 *Art: The Belief in Primacy of Living Consciousness*

First and foremost, the belief in primacy of living consciousness revealed itself in experimenting with general properties of phenomena. In visual art, some cases of such experimenting can be found in palaeolithic cave paintings, which, along with magnificent realistic images, contain images that combine features of people and animals. Other instances of experimenting with the raw phenomena's general properties can be observed throughout history, from Egyptian to Medieval arts [4, 5]. However, these concessions of traditional art to epistemological scepticism were exceptional and confirmed rather than denied the art's adherence to basic properties of raw phenomena, such as identity and permanence. Academism and other versions of realistic art observed 3D geometrical perspective, arranged objects in accord with the law of gravitation and represented physical time by portraying ruins and elderly people. The tendency towards primacy of living consciousness became taking momentum in the middle of the nineteenth century, with the emergence of new trends in visual arts.

The message that these trends brought was that all phenomena are *linked by some deep level unity*. This message is evident in impressionism, cubism, optical art, and other trends. The Rouen Cathedral's series by Oscar-Claude Monet portrays a gothic cathedral dissolving in the morning air [6], in Pablo Picasso's painting 'Portrait of Ambroise Vollard' a person's face is emerging from the elements of the environment as if called into existence by a magical spell [7], and graphic works by Mauritz Escher, such as 'Sky and Water', show gradual transformation of one object into another [8]. By removing the object\environment and object\object boundaries, the artists challenged the phenomena's identity.

Following impressionism, artists began to explore visual phenomena by distorting physical space (e.g., the 'double face' portraits by Pablo Picasso [9]), defying physical gravitation (e.g., the 'Waterfall' by Mauritz Escher [10], the 'Flying Lovers' by Marc Shagall [11]), 'deconstructing' space through spilling paint on canvas (e.g., Jackson Pollock's abstract impressionism [12]) and creating the arbitrary combinations of lines and shapes (e.g., Vassily Kandinsky's abstract art [13]). By suspending the intrinsic properties of phenomena visual arts opened the door into the depths of 'possible universes', trying to imagine how these possible worlds might look. And here they came—the magical universes by Giorgio de Chirico, Rene Magritte, and Salvador Dali. The visual field flows, objects transform one into another, living and non-living entities merge in one. The reality melts, solidifies, and melts again. Phenomena discontinue their links with their RCs and go into a free flight.

A similar relaxation of intrinsic properties of raw phenomena can be seen in literature. The great realistic literature of the nineteenth–early twentieth centuries, represented by Honoré de Balzac (1799–1850), Leo Tolstoy (1828–1910), and Theodore Dreiser (1871–1945), rested on the assumptions of the existence of absolute truth and objective view of events. In the beginning of the twentieth century these assumptions began to crumble; instead of bringing reality to a single point of view, writers switched to reporting their 'stream of consciousness'. In the novels by Marcel Proust (1871–1922), James Joyce (1882–1941), and Franz Kafka (1883–1924), just like in visual arts, the mixture of realistic and fantastical events led to the dissolution of boundaries between ordinary and magical realities. Finally, postmodernist writers declared the 'relativity' of truth, thus replacing the objective description by subjectively biased narratives [14]. The reality lost its 'backbone'—the objective truth—and turned into the 'literary invertebrate' in the form of a discourse. Following the experiments in visual arts, modern writers opened the door into the unconscious living consciousness.

In its silent dispute with science, art asserts the idea of primacy of living consciousness over science generated reality. In the minimalist art, objects are stripped of their colours and textures [15]. Bare cubes, squares, and other shapes carry a hidden message saying that *RCs become pointless when torn off their phenomenal foundation*. Another way to deliver this message is designing phenomenal objects whose RCs cannot be built, such as the 2-dimnsional images of impossible objects by the Swedish artist Oscar Reutersvärd (1915–2002) [16] and by the Dutch artist Mauritz Escher (1898–1972) [17]. Finally, one more way of demonstrating the creative power of living consciousness was the art of pointillism—a technique of the impressionist art that uses the ability of the eye to fuse colour spots into a full range of tones [18]; looking at paintings by Georges Seurat (1859–1891) and Paul Signac (1863–1935) we see not a mixture of coloured dots but a palette of fresh colours and objects instead.

Art was at the forefront of embarking on the belief of the primacy of living consciousness, but this belief influenced other practices as well. In psychology, this belief brought about trends such as associationism, psychoanalysis, and logotherapy by Viktor Frankl [19]. In experimental psychology, there appeared projective techniques (e.g., TAT, Rorschach, the method of free associations), studies of anomalistic phenomena (e.g., remote viewing and psychokinesis [20, 21]), and new trends in studying living consciousness, such as the study of magical thinking [22, 23]. Even in physical science the idea of primacy of consciousness was employed in theories suggesting that observation plays a crucial role in defining the properties of a quantum entity [24, 25]. One more trend in interpreting reality as a product of living consciousness emerged in the history of science, when reality became to be interpreted as socially and culturally constructed. The most influential case in this trend was the idea that the scientific view of the world is based not on objective reality but on the 'paradigm'—the prevailing beliefs in a scientific community, whereas objective reality is given a secondary role of 'accumulation of anomalies' that don't fit the accepted paradigm [26]. In this trend of exploration, the collective rather than individual living consciousness was employed, in the form of culturally and historically embedded beliefs (see Chapters 9 and 10 for the elaboration on the collective living consciousness).

6.2.4 Science: The Belief in Primacy of Rational Constructions

6.2.4.1 Informational Reductionism

Information reductionism claims that science-generated objects not only influence consciousness but generate it. In neuroscience, living consciousness is often interpreted as an epiphenomenon—a side effect of neural processes in the individual's brain that doesn't have causal powers over the individual's behaviour (see Chapter 4 for more on that).

A similar attempt to reduce living consciousness to RCs comes from mathematics. According to the proponents of the Integrated Information Theory (IIT), living consciousness of a certain system is a result of the exchange of information between this system's subsystems [27]. The system could be our brain, and subsystems—different groups of neurons or different regions in the brain. The system could also be the groups of atoms, molecules, or even numbers. The degree of any subsystem's output can be assessed mathematically as a value called 'phi', and this value assesses the degree of the system's living consciousness. This approach implies that living consciousness may be a property not only humans and other living organisms but of inanimate systems as well. By doing so the IIT makes a pact with panpsychism. However, like any other reductionist theory, the IIT runs into a logical circle. Indeed, the IIT begins with identifying the fundamental properties of living consciousness—existence, composition, information, integration, and exclusion—which are then used to define the physical substrate of living consciousness—simple systems, such as logic gates or neuron-like elements. Next, the theory postulates that the physical substrate can form complexes of a higher order that develop into the aforementioned fundamental properties of living consciousness, thus locking the argument into a circle.

6.2.4.2 Evolutionary Reductionism

According to one version of evolutionary reductionism, our living consciousness is phenomenalistic reality adapted by evolution to help organisms to survive [28]. As a result, the living consciousness functions like a computer interface, in which 'icons'—evolutionary adapted phenomenalistic reality—are linked to 'folders'—the true reality out there. In this approach the living consciousness is a system of 'useful illusions', whereas the 'true reality' is a mystery which evolution made forever hidden from view. In a sense, this form of reductionism is a modified version of Kantian division between phenomena and noumena, in which phenomena are viewed as 'evolutionary adapted' noumena.

Adjacent to the previous version, another version of evolutionary reductionism is using Darvin's theory as a universal retrospective explanatory framework for living consciousness [29]. For example, from the observation that modern humans prefer pictures of faces with the limbal ring to pictures of faces without the limbal ring scientists infer that our distant ancestors used to find partners with this feature more attractive in the past, because sexual partners with such feature were healthier and offered a better fitness payoff than partners without the feature [30]. This way of reasoning implies a reversal of the causal sequence: constructing the evolutionary past on the basis of observations over behaviour of modern humans turns into explaining properties of living consciousness of modern humans by the events in the past. The reversal of causal sequence makes it possible to explain any feature of modern people's living consciousness by inventing a plausible 'evolutionary story'. In other words, evolutionary reductionism is not an explanation of modern phenomena with their evolutionary origins, but a construction of the past on the basis of observations made in the present.

6.2.5 Consequences of Believing in the Primacy of Consciousness and in the Primacy of Rational Constructions

To summarize, the epistemological scepticism proved to be a mixed blessing for modern culture. Among its positive consequences, along with the aforementioned new trends in art, one can list the development of projective methods for the study of the living consciousness and different forms of existential psychotherapy. More recently, there appeared new trends in the experimental psychology of living consciousness such as the study of magical thinking. However, there were also negative consequences, such as polysemy and arbitrariness in the interpretation of phenomena of living consciousness, and fragmentation of studies of consciousness into opposing schools and currents (see Chapter 1 for more on that).

A similar two-edged sword was the epistemological reductionism. Positive consequences of the belief in primacy of RCs were developing new and more subtle methods of studying the relationship between the cerebral cortex and mental phenomena, such as Magnetic Resonance Imaging (MRI), and creating a detailed map of brain localization of mental functions. Negative consequence was that living consciousness lost its relative autonomy and became viewed as an epiphenomenon—a product of physical processes casted on the inner screen of consciousness which doesn't have any causal powers in the physical world.

6.3 Bifurcation 2: Symbolism

The second bifurcation strikes the type of representation of living consciousness in the domain of objectified consciousness—symbolism. Since the Age of Enlightenment, living consciousness has been objectified, stored, and transmitted in the form of symbols—literal texts or schematic images. When reading a text, a person recorded a sequence of symbols into the so-called 'analogues'—mental or visual images in his or her living consciousness similar to real perceived objects, manipulated them internally, and proceeded to actions, or decoded the results of thinking back into texts and diagrams of the 'hard' objectified consciousness.

Today, the basic *symbolic type* of representation has been transformed into the *digital type*, and the *animated type* has spun off from the digital type of representation (a postmodern phenomenon) (Fig. 6.3).

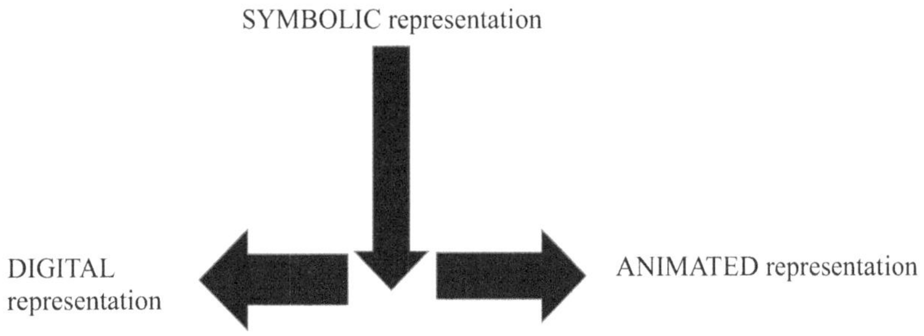

Fig. 6.3 Bifurcation 2: Symbolism. The basic SYMBOLIC type of representation of living consciousness bifurcated into DIGITAL and the ANIMATED types

6.3.1 Digital and Animated Types of Representation

Digital representation finds its extreme form in programming languages, in which symbols (e.g., letters and numbers) are replaced by digital codes—bytes (see Chapter 4 for more on this). This development, coupled with the invention of computers, made it possible to convert symbolic texts into animated pictures. Psychologically, working with animated pictures is easier than working with their symbolic originals, because symbolic representation is not commensurable with living consciousness, whereas the animated representation is. For example, the symbolic phenomenon of the word FISH (i.e., a piece of objectified consciousness) doesn't look like the perceptual image of the creature we are seeing in the pond or imagining in the mind, whereas the animated image of a fish on a computer screen does. Consequently, the cognitive appropriation of symbolically presented information, which takes place in the buffer zone of the internal objectified consciousness, happens through *decoding*—converting messages from the symbolic into the analogue form (see Fig. 6.4, rationality-logical route). In the process of decoding another important process is taking place—the assessment of the incoming messages on their truthfulness and reliability (see Chapter 7 for more on that).

In contrast, consuming messages presented in the form of animated images happens through *magical participation*—the identification of living consciousness with the external images, which is the basic mechanism of living consciousness. Since animated imagery is commensurable with the imagery of living consciousness, a piece of animated objectified consciousness (e.g., the image of a moving fish on a computer screen) gets into the living consciousness directly, bypassing the buffer zone (see Fig. 6.4, participatory-magical route) .

All and all, the tendency that prevailed in the last decades of the twentieth century and in the twenty-first century has been that people prefer to cognitively appropriate information coded in the form of images rather than symbols. As a result, the type of contact of the person's living consciousness with 'hard' objectified consciousness turns from rationality-logical into participatory-magical.

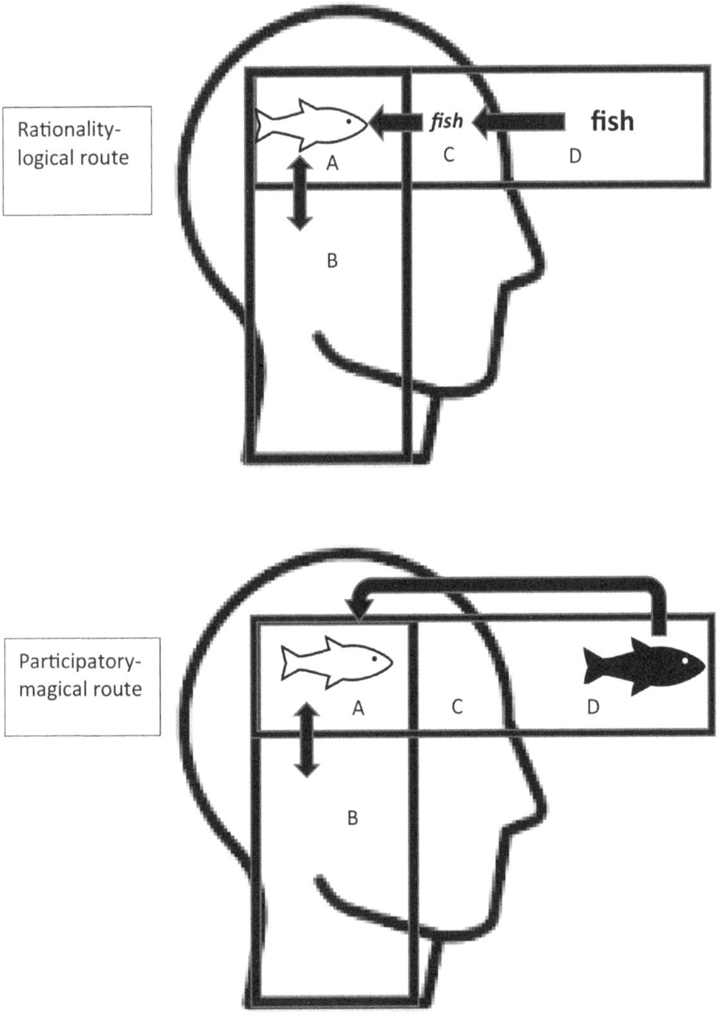

Fig. 6.4 Rationality-logical route of objectified consciousness to living consciousness (objects' representations in objectified and living consciousness are incommensurable) and participatory-magical route (objects' representations in objectified and living consciousness are commensurable). **A** Conscious living consciousness, **B** Subconscious living consciousness, **C** Internal objectified consciousness (the buffer zone), and **D** External objectified consciousness

6.3.2 *Consequences of the Transition from Symbolic to Digital and Animated Types of Representation*

Positive consequences of the transition from symbolic to digital type of representation of living consciousness included creation of effective means of rapid processing and transmission of information, and as a result—development of artificial intelligence (AI). Among negative consequences was the illusion that it is possible to digitize not only images of perception and logical thinking, but also creative thinking, emotions, and feelings. Another negative consequence was the 'Copernican revolution' in the relationships between living and objectified consciousness, when objectified consciousness in the form of SGO's began to be viewed as a cause of the living consciousness, and the living consciousness turned into an epiphenomenon (see Chapter 4 for more on this).

As for the consequences of transition from digital to animated type of representation, positive consequence was that assimilation of information stored in the form of images became faster and more economical than assimilation of information stored in the form of symbols. Another positive consequence was that exposure to animated magical reality can enhance creativity. Creativity is the production of novelty literally from nothing, and this is one of properties of living consciousness. Unlike written texts, animated imagery is a type of reality commensurable with the imagery of living consciousness. As a result, watching animated magical events 'resonates' with living consciousness and stimulates divergent creativity of living consciousness (see [31] and Chapter 2 for more on that).

However, like it was the case with transition to digital representation, positive consequences of the transition to animated representation were accompanied with negative ones. Unlike a symbolic text, a visual image directly affects emotions. Consequently, a visually presented message bypasses critical thinking and its effect on the living consciousness is faster and stronger than the effect of a textual message. As a result, manipulation of people's living consciousness in the domain of politics and commerce through animated images (movies, television, computer games) becomes easier, and it becomes difficult to distinguish fakes from real events (see Chapter 8 for more on that). The presence of the participatory-magical route towards living consciousness opens to fraudsters the opportunity to smuggle their messages directly to our living consciousness—the phenomenon known as suggestion (see Chapter 7 for the elaboration on this phenomenon).

Another negative effect of storing living consciousness in the form of animated images was in the domain of education. Because the cognitive load of processing information presented in the animated form is lower than the load required for processing information given in the symbolic form, most children today prefer watching and listening to reading. Since not all complex notions can be converted into images, the 'reading to watching shift' leads to the simplification of thinking and the prevalence of suggestion over rational thinking in education. In addition, symbolic language is a carrier of logical thinking, and logical thinking is necessary for conversion of creative production of living consciousness into objectified logical structures that can be communicated to other people. In other words, the transition from symbolic to animated form of objectified consciousness brings about the gain in creativity but also the weakening of the ability to convert the results into a transferable form. As the shift from symbolic to animated form of information processing is non-reversible, the optimal strategy in education could be maintaining a balance between presenting educational material in animated and symbolic forms.

6.4 Bifurcation 3: Limited Freedom

Finally, the third and most important bifurcation strikes our belief in the limited freedom. Morality and jurisprudence of the age of Enlightenment and modern era were based on the belief that our free will is limited by demands of society and environment. A person is responsible for the consequences of his or her actions, but he or she is not responsible for the environment in which he or she was born, to which gender and nation he or she belongs. In addition, the age of Enlightenment, which in Western Europe reached its peak in the seventeenth and eighteenth centuries, stood close to the beginning of the Reformation, and people's belief in God were still strong. Both Catholics and a Protestants of that epoch felt that their actions on Earth are being observed by God, and this put limitations on people's freedom to commit actions that contradicted the demands of religion and morality.

In the twentieth century, the belief in the limited freedom bifurcated into the belief in unlimited freedom of man (the postmodern phenomenon) and the belief in hard determinism of nature and society over the individual (Fig. 6.5).

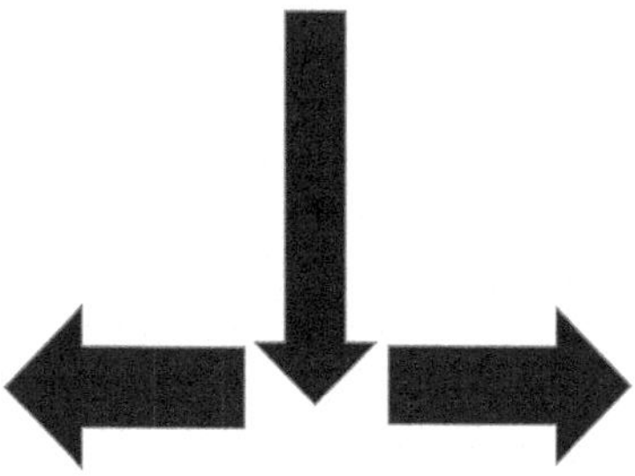

Fig. 6.5 Bifurcation 3: Freedom. The basic belief in LIMITED freedom bifurcated into the belief in UNLIMITED freedom, and in HARD determinism

6.4.1 Belief in Unlimited Freedom

Today, religion declined in its social and cultural significance. This doesn't mean that most people became atheists, but even in believers in God the feeling that their every action is under the divine supervision is weakened or disappeared. In his book 'Thus Spoke Zarathustra', published in 1883, German philosopher Friedrich Nietzsche boldly claimed that 'God is dead'. In the novel 'Brothers Karamazov' (1880) by Russian author Fyodor Dostoevsky one of the characters famously said, "If there is no God, everything is permitted". In 1843, German economist and philosopher Karl Marx published his work 'Contribution to the Critique of Hegel's Philosophy of Right', in which he called religion 'the opium of the people'. Coupled with the great achievements of science and medicine, which substantially decreased peoples' dependence of the uncontrolled forces of nature such as hunger and epidemies, the decline of the belief in divine retribution for sins in the earthly life made people feel that their freedom of satisfying their secret desires, including the traditional vices, is unlimited. Theoretically, this feeling culminated in philosophical trend of existentialism. According to existentialist theories human freedom is unlimited. A person chooses everything—his or her actions, morality, environment, nation, gender, and even life itself [32]. The French existentialist philosopher Albert Camus wrote "There is but one truly serious philosophical problem, and that is suicide" [33, p. 3].

6.4.2 Belief in Hard Determinism

In the present time, some scientists consider a person as a system entirely determined by the information recorded on the DNA molecule, while others consider a person to be an avatar living in a computer simulation (see Chapter 5 for the elaboration on that). Physical determinism is echoed by social determinism. With the advent of mass media, a new phenomenon arose—a totalitarian media narrative. Totalitarian media narratives are social myths that feed on suggestibility of living consciousness of individuals. Captured by the totalitarian media narrative, a person becomes inaccessible to logical arguments and takes a one-sided abstract view on what is happening in the world. By blocking fair-minded thinking totalitarian media narratives disturb the balance between rational and magical elements withing the living consciousness. Totalitarian media narrative is insensitive to political structure of society and can equally zombify citizens of both authoritarian and democratic societies. A person's identification with totalitarian media narratives makes the person insensitive to the unconditional value of an individual human life and to the life-saving 6th commandment 'Though shalt not kill' (Exodus 20:13) (see Chapter 9 for more on totalitarian media narratives).

6.4.3 Consequences of Believing in Unlimited Freedom and Hard Determinism

Positive consequence of the belief in unlimited freedom was a shift from orientation towards social or divine punishment for disobedience to orientation towards 'moral autonomy'—the internal desire to maintain one's moral self-esteem. Most prominently this change was expressed by Immanuel Kant in his 'Groundwork of the Metaphysics of Morals' (1785), in which Kant formulated his categorical imperative of morality "Act only in accordance with that maxim through which you can at the same time will that it become a universal law". Another positive consequence was the emergence of social movements for the rights of racial, national, and sexual minorities. Negative consequences were the emergence of politically motivated terrorism and the danger of devaluating the traditional Christian values by turning sexual minorities into politically influential groups, which could potentially lead to underpopulation of entire nations. Another negative consequence is the cases of suicidal revenge that recently spread in some industrial counties, particularly hitting teenagers. When a non-believer in the divine retribution makes a decision to take his or her own life, the person suddenly realizes that he or she is free to take other people's lives as well without consequences for himself or herself. Therefrom there appeared seemingly unexplainable cases of mass killings at schools and other people's gatherings by teenagers and adults who thus far looked like quiet and ordinary individuals.

As for the belief in hard determinism, its positive consequences in biology are genetic engineering, and in psychology reduction of feelings of anxiety and uncertainty caused by the need to make choices. Negative consequences in biology are artificially created genetic deformities, and in psychology orientation towards personal interests rather than towards moral values. In the domain of social life, negative consequences were the blockade of critical thinking, manipulation of people's living consciousness through totalitarian media narratives, and the phenomenon earlier described as the 'escape from freedom' [34].

6.5 Global Results of the Bifurcation of Living Consciousness: Conclusion

The Covid-19 epidemic restricted our movements around the world and increased the feeling of the loss of personal freedom. The global confrontation between Russia and the West that is taking place today tears the world in parts and throws a person into the maelstrom of totalitarian media narratives. A person or a country with whom we were friends yesterday could turn into an anger-filled opponent, ready for insults, threats, and even outright violence. The suspension of the fear of divine retribution made nonbelievers in God free to commit immoral and antisocial actions in absence of external control, whereas fanatical believers in God today, like in the ancient times, may commit the acts of terror to win the after-life rewards. On the level of nations and states, the gap between the technical possibilities of achieving personal gains and the orientation towards moral values has reached an explosive level (see Chapters 9 and 10 for more on that). The question of what these bifurcations in our living consciousness will lead to in the future remains open. Perhaps, the positive consequences of these bifurcations will prevail and the humanity will move to a new stage of development, becoming more free, creative, and moral. It is also possible that the grave repercussions of the bifurcations will take an upper hand and the humanity will perish in a nuclear war, fall into the chaos of the New Middle Ages, or degenerate into a species of emotionless cyborgs.

All and all, changes in our understanding of living consciousness that followed the epoch of modernity, along with achievements in science and technology bring about losses in the spiritual domain and controversial phenomena in modern life, some of which will be analysed in Part III of this book.

References

1 Subbotsky, E. (2023). *Examining the psychological foundations of science and morality: Explaining the inexplicable.* Routledge.
2 Hicks, S. (2011). *Explaining postmodernism: Scepticism and socialism from Rousseau to Foucault.* Ockham's Razor Publ.
3 https://plato.stanford.edu/entries/postmodernism/
4 https://www.britannica.com/art/Egyptian-art
5 https://www.anothermanmag.com/life-culture/10393/hieronymus-bosch-s-monsters-explained
6 https://en.wikipedia.org/wiki/Rouen_Cathedral_(Monet_series)
7 https://www.pablopicasso.org/portrait-of-ambroise-vollard.jsp
8 https://www.artsy.net/artwork/m-c-escher-sky-and-water-i
9 https://no.pinterest.com/pin/354588170662330879/
10 https://en.wikipedia.org/wiki/Waterfall_(M._C._Escher)
11 https://www.dailyartmagazine.com/bella-and-marc-chagall/
12 https://www.christies.com/lot/lot-4288617/?intObjectID=4288617
13 Lindsay, K., & Vergo, P. (1994). *Kandinsky: Complete writings on art.* Da Capo Press. ISBN 9780306805707.
14 Kakutani, M. (2018). *The death of truth.* William Collins.
15 https://www.britannica.com/art/Minimalism
16 https://www.illusionsindex.org/i/impossible-triangle
17 Ernst, B. (1978). *The magic mirror of M.C. Escher.* Taschen.
18 Block, J., & Lee, E. W. (Eds.). (2014). *The neo-impressionist portrait, 1886–1904.* Yale University Press.
19 Frankl, V. (2006). *Man's search for meaning: An introduction to logotherapy.* Beacon Press.
20 Bem, D. J., & Honorton, C. (1994). Does Psi exist? Replicable evidence for an anomalous process of information transfer. *Psychological Bulletin, 115,* 4–18.

21 Dunne, B. J., Nelson, R. D., & Jahn, R. G. (1988). Operator-related anomalies in random mechanical cascade. *Journal of Scientific Exploration, 2*, 155–179.
22 Nemeroff, C., & Rozin, P. (2000). The making of the magical mind: The nature and function of sympathetic magical thinking. In K. S. Rosengren, C. N. Johnson, & P. L. Harris (Eds.), *Imagining the impossible: Magical, scientific, and religious thinking in children* (pp. 1–34). Cambridge University Press.
23 Subbotsky, E. (2010). *Magic and the mind: Mechanisms, functions and development of magical thinking and behaviour*. Oxford University Press.
24 Faye, J. (2019). *Copenhagen interpretation of quantum mechanics*. Stanford Encyclopedia of Philosophy. https://plato.stanford.edu/entries/qm-copenhagen/#Bib
25 Wheeler, J. A. (1990). Information, physics, quantum: The search for links. In W. Zurek (Ed.), *Complexity, entropy, and the physics of information*. Addison-Wesley.
26 Kuhn, S. T. (1996). *The structure of scientific revolutions*. University of Chicago Press.
27 Tononi, G. (2015). Integrated information theory. *Scholarpedia, 10*(1), 4164. http://www.scholarpedia.org/article/Integrated_information_theory
28 Hoffman, D. D. (2020). *The case against reality*. Penguin Books.
29 Hawking, S. W., & Hertog, T. (2006). Populating the landscape: A top-down approach. *Physical Review*. https://arxiv.org/pdf/hep-th/0602091.pdf
30 Peshek, D., Sammak-Nejad, N., Hoffman, D. D., & Foley, P. (2011). Preliminary evidence that the limbal ring influences facial attractiveness. *Evolutionary Psychology, 9*, 137–146.
31 Subbotsky, E., Hysted, C., & Jones, N. (2009). Watching films with magical content facilitates creativity in children. *Perceptual and Motor Skills, 111*, 261–277.
32 Sartre, J. P. (2007). *Existentialism is a humanism*. Yale University Press.
33 Camus, A. (1955). *The Myth of Sisyphus*. https://people.brandeis.edu/~teuber/Albert_Camus_The_Myth_of_Sisyphus_Complete_Text_.pdf
34 Fromm, E. (1994). *Escape from freedom*. Holt.

PART III

Living Consciousness in Modern Life and Society

CHAPTER 7

Reason in the Cobweb: Magical Manipulation of Living Consciousness

7.1 'Social engineering' in the Context of Criminology

Today, a new type of crime is spreading rapidly in large cities of the world. Typically, crimes are committed overtly, and their mechanisms are psychology transparent. From time immemorial, people have been robbed, deceived, and hypnotized. In such cases, the victims of the crimes acted rationally; after all, when a gun is pointed at you, it is reasonable to give up a wallet than one's life or health, and if you believe a clever fraudster who is skilfully deceiving you, then your actions seem reasonable to you. As far as hypnosis is concerned, a person is practically turned into a zombie on the basis of the laws of physiology rather than psychology. But consider, for example, this real life case.

An elderly woman—a pensioner and an educated person, being alone in her apartment, receives a telephone call from an 'employee of a pension fund' (also a woman), with an offer to pay a visit to discuss a range of free social services. Even though a few days before the event a district police officer personally had warned the pensioner that scammers were operating in the area, the woman lets the 'employee' into her apartment and quite consciously and voluntarily gives away all the cash and family jewellery. Only after the departure of the so-called employee, the pensioner begins to doubt her actions and calls the district police officer. But it is too late. The question arises why an educated person in the normal state of mind,

E. Subbotsky, *The Magic of Living Consciousness*,
https://doi.org/10.1007/978-3-031-54708-9_7

being alone in her apartment, allowed a stranger into her home, despite she had been warned about the danger of fraud lurking in the area and could have called the local policeman or her relatives for an advice? And why the person in full consciousness voluntarily led the fraudster to her money despite earlier she didn't display the whereabouts of her money even to close relatives? At that, there was no threat to her life and health, she was not hypnotized, and the deception applied to her was sewn in white thread. Indeed, the representative of the pension fund had a reason to ask the client to show her passport or a pension certificate, but not to display her cash and family heirlooms. Why then did the victim fail to see the contradiction between the role of the 'civil servant' and the servant's actions? And if she did see the contradictions, why did she fail to use her critical judgement and led the fraudsters to whereabouts of her money? There is some psychological mystery in this.

It is obvious that in this situation the victim acted consciously, but her *reason* was completely or partially turned off. The word 'reason' is defined as 'the capacity of applying logic consciously by drawing conclusions from new or existing information, with the aim of seeking the truth' [1]. Since reasonable thinking applies logic, it is a part of objectified consciousness, which means that living consciousness and reason are not the same. You can act consciously, but unreasonably. The question is why in the described case the human reason is, as it were, entangled in a cobweb and cannot influence the conscious actions of a person, only powerlessly 'moving its paws'. In this chapter, we will try to give the answer to this question based on the nature of living consciousness.

7.2 Magical Participation as a Psychological Mechanism of Living Consciousness

As argued in Chapter 3, living consciousness functions in accord with the laws of magic. According to one of these laws, a person's word or intention can by itself, without any expenditure of physical energy, affect physical reality (the law of 'mind over matter'). For thousands of years people believed in the magic of words and tried to control the forces of nature and other people with the help of magic spells. It can be assumed that in the field of communication the word was originally not a concept, but a stimulus that prompted a listener to action. Today most educated people consciously don't believe in magic. As a result, magical thinking separated from magical beliefs: whereas in the domain of the imagination and art we still consciously practice magical laws, in the domain of our everyday practice with physical objects and communication with people we rely on the laws of nature and formal logic. Nevertheless, psychological studies show that subconsciously many rational people today still hold the belief in magic of words. Most clearly this subconscious belief reveals itself in the domain of communication [2].

The subconscious belief that a sound signal or a spoken word (magic spell or verbal order) is identical with reality unites people with animals. For an antelope, the sound signal of another antelope about the approach of a predator is not a message that can be critically evaluated and verified for truth but the real presence of danger and incentive for an immediate action. Unlike animals, humans developed the ability of rational (that is, guided by reason) thinking, in which a person separates a verbal statement from reality, considering the statement from the 3rd person's perspective. From an order to an immediate action a verbal statement turned into a model of reality that can be either true or false.

As a result, the consciousness of a modern person is split into two domains. In one domain, we have an inner world of living consciousness working in accord with the laws of magic (see Chapters 1 and 3 for more on that). In this world—the world of magical thinking and magical experiences—the stiff boundary between thinking and feelings does not exist. Feelings are involuntary, they arise and disappear by themselves, they are connected not by physical forces or formal logic but by magical participation. Feelings are experienced, not judged. But along with the 'person-experiencer' in our mind lives another person—'the person-observer'. The 'person observer' dwells in the domain of reason—the domain of internal objectified consciousness (see Fig. 1.1). When we are in the position of the 'experiencer', we perceive the statement addressed to us as an order, but in the position of the 'observer' we critically evaluate the statement and can accept or reject it. Under normal circumstances, the 'person-observer' has an upper hand over the 'person-experiencer' in controlling our actions. A message or a request from other people goes to the 'person-observer', is evaluated by him or her in terms of its truth or usefulness, and then either rejected if it is unprofitable or illogical, or forwarded to the 'person-experiencer' who acts in accordance with the content of the message or request. It means that messages coming from other people, in order to reach our living consciousness, must follow the successive-logical route and go through the buffer zone of the internal objectified consciousness, in which the messages are 'distilled' through the filter of reason and assessed on their truthfulness and reliability. It could also happen that we deactivate our 'person-observer' and the message proceeds via the participatory-magical route directly to our 'person-experiencer' (see Fig. 6.4). Usually, this deliberate deactivation of the buffer zone of our critical thinking happens when we are involved into communication with magical reality such as perception of art, or in communication with friends and family when we rely on the principle 'not to judge but to understand'. However, there could be one more way for a message to get to our living consciousness—the 'message smuggling' way. When a message is smuggled to our 'person experiencer', the message goes via the buffer zone of critical assessment but the reason is in the state of relaxation and unable to stop the message from coming through. This explains the phenomenon of 'reason in the cobweb': we hear or see the message, we understand that the message is wrong, but still follow the order or suggestion that this message contains.

The easiest way for a producer of a message to smuggle the message to a person's living consciousness is when the producer and the receiver of the message are the same person. For example, a person (a 'producer') thought about the possibility of buying a car, and there are funds available, but the person's 'observer' says: "Don't buy, because your driving skills are imperfect, there are traffic jams on the streets, there is nowhere to park the car, keeping and servicing the car is expensive, public transport is easier and faster, and for elderly people like you public transfer is free". But the person's own 'experiencer' is nearby, so the message (the idea of buying a car) can ignore the 'observer's warnings and find its way through the buffer zone straight to the 'experiencer', where the message turns into a desire—a stimulus to action. The person's 'smuggler' defeats the person's 'observer' but soon begins to feel that he or she made a mistake. After a couple of years of torment on the crowded roads, the person sells the car at a big loss. In the end, reason won, but at what cost! The same thing can happen to us when a message comes from other people, when our 'experiencer's belief in the magical power of the message overpowers the 'observer's critical thinking and we deliberately accept overtly ridiculous proposals. The question arises what psychological basis is hidden behind the 'message smuggling' rout to our living consciousness?

In ordinary speech, the 'message smuggling' phenomenon is known as *suggestibility*, but its psychological mechanism became clear only with the proof of the presence in our living consciousness of *magical participation in communication (PC)*—the identification of the spoken message with reality contrary to the understanding that the message is wrong. This unconscious belief in the 'word/reality identity' is the 'psychological gunpowder' on which the 'spark' of the suggestive message falls, thus triggering the PC. Without the 'spark' the 'gunpowder' will not explode, and without the 'gunpowder' the 'spark' will get extinguished. Whereas the belief in the word/reality identity is the *psychological basis* of the PC, the *behavioural manifestation* of the PC is the *dissociation between reasoning and action*. On the contrary, the behavioural manifestation of rational behaviour is the *harmony between reasoning and action*. A person behaves rationally when his or her actions are guided by a critical assessment of the message addressed to him or her.

The PC differs both from emotionally conditioned behaviour and from behaviour caused by hypnotic trance. Emotions distort people's critical reasoning, and in the state of hypnotic trance critical reasoning is completely disabled, whereas under the PC the ability of critical reasoning is present but relaxed and unable to guide actions. Accordingly, when the PC is activated our unconscious belief in the word/reality identity forces us to do what the reason forbids.

But not every suggestive message takes control over our behaviour. We regularly hear and read various false and unfavourable messages and requests addressed to us, but don't view these messages as an incentive to action. For the 'spark' to break through the criticism of the 'observer' and get to the 'gunpowder' of the unconscious belief in the word/reality identity, the spark must be strong enough. To provide a sufficiently powerful spark, the message must come from a person or organization in whose authority the receiver of the message believes. Thus, in traditional societies people used to seek magical healing not from any tribesman but from a sorcerer or a healer [3]. When we are suggestible, we perceive the message as if it were spoken by God, and God never errs. For a small child, an adult looks almost like God, so children are especially vulnerable to the PC; psychological experiments demonstrated that many 4- to 6-year-old children imitate even those actions of adults that they openly call erroneous [4]. A statement that adults perceive as a joke may be perceived by children as a reality. It is no coincidence that law officers who investigate cases of 'social engineering' often notice that victims of this kind of fraud 'become like children'.

Another condition of the PC activation is addressing the sufficiently powerful 'gunpowder' within our living consciousness—some image which is personally important. If a suggested message addresses something unimportant in our life, it is unlikely that the message could ignite the belief in word/reality identity. It is one thing to say that a magic spell would affect a life of a stranger and quite another to say that the message would affect your life or life of your close ones.

To examine the presence of the PC in modern educated adults, an experiment was conducted [5]. In a pre-test interview, participants (university graduates and postgraduates) were tested on their understanding of the difference between proper (supernatural) magic and stage magic. The participants who passed the test and said they didn't believe in magic took part in two experimental conditions. In one condition (personal involvement with magic) participants were asked to imagine that a witch approached them on a dark street and offered to cast a magical spell that would change their future lives. In one case, the spell was supposed to make the participants' life good and trouble-free (a desirable outcome), and in the other to make them servants of a dark supernatural power (an undesirable outcome). Participants were asked if they would allow the witch to cast the spells on their future lives.

It was expected that if participants didn't succumb to the PC (identification of the message with reality), they would treat the witch's proposal as a fantasy and accept or reject both kinds of spell in a ratio of about 50×50. However, if participants experienced the effect of the PC, then they would unconsciously believe in that the magic spell could affect their future lives. In this case, about half of the participants would accept the spell with the desirable outcome, but most would reject the spell with the undesirable outcome.

In another condition (no personal involvement with magic) participants were asked what answer they would recommend to the plot character (a scientist who does not believe in magic) if the character were in the imaginary situation with the witch. Since in this situation the participants' lives were not at stake, participants rated magical suggestions from a 3rd person perspective. As a result, the split between the unconscious belief and the consciously proclaimed disbelief in the word/reality identity was eliminated: the participants' latent fear that the spell might affect their future lives no longer contradicted their consciously declared disbelief in that the magic words could affect reality, since the witch's message was aimed at the stranges' life rather than the participants' future lives.

The results confirmed the existence of the PC: in the personal involvement condition about half of participants accepted the good spell, but almost all rejected the bad one and justified their decision by their worry that the bad spell might work on their future lives. However, in the no personal involvement condition, about half of the participants recommended the character of the story to accept both spells. When these participants were then asked what they would say to the witch if the offer were made to them personally, the distribution of the participants accepting and rejecting the spells became as in the personal involvement condition.

In the experiment described above an imaginary witch was the 'authority' who ignited the 'gunpowder' of the unconscious belief in the word's magical powers. In reality, the authority could be any individual who has a permanent or temporary power over a person, even a psychologist who is conducting an experiment. Thus, in a notable experiment by the American psychologist Stanley Milgram adult participants were offered to teach other participants certain skills and punish them with an electric shock for unsuccessful attempts [6]. When the 'student' erred repeatedly, the 'teacher' was instructed to increase the voltage. In one of the conditions of this experiment, 62% of the 'teachers' followed the experimenter's instruction and increased the voltage to a near-lethal level of 450 volts. The 'teachers' did not know that in reality the 'students' did not receive electric shocks but only imitated the feeling of pain. It was obvious from the 'teachers' behaviour that they were aware of that the experimenter's order to increase the voltage was morally wrong: they exhibited signs of anxiety, argued with the experimenter, but continued to increase the voltage. In Milgram's own account, "something akin to fields of force, diminishing in effectiveness with increasing psychological distance from their source, have a controlling effect on the subject's performance" [ibid, p. 147]. However, the psychological nature of this 'fields of force' remained unclear.

What prevented the 'teachers' from simply stopping the experiment and disobeying the experimenter's demand? Since the 'teachers' could withdraw from the experiment at any time without a loss of a promised reward, the only plausible explanation for their reluctance to stop increasing the voltage was their unconscious belief in that the experimenter's demand was 'an order from God' that couldn't be violated, just like the order couldn't be violated by the biblical character Abraham when God ordered him to sacrifice his beloved son (Gen. 22:1–19). This was a clear case of the PC, based on the participants' unconscious belief in the authority's special powers. "But what are the origins of this belief?"—the reader might ask.

7.3 Participation in Communication in a Historical Frame

The answer may be hidden in the prehistory of the mind. Let's assume that at the dawn of mankind people began to obey verbal commands of their leaders not as a submission to brute force, but because they identified leaders with the divine forces. In the animal kingdom, a dominant male has to constantly fight for his superiority and loses this position with age, but in human societies kings' right to rule is sanctioned by God. Unlike obedience to brute force, obedience to God is unconditional, since the word of God is identical with reality. In the modern time secular sources of power (e.g., doctors, priests, and political leaders) became accessible to people's critical thinking. However, the unconscious identification of secular authorities with divine powers remains a part of living consciousness in modern people. This unconscious identification provides a psychological explanation for otherwise mysterious social phenomena. One of these social phenomena is the PC revealed in the reviewed psychological experiments.

In principle, any verbal message that we send to other people with the aim of making their behaviour fit our goals is an attempt to intervene in other people's living consciousness, with the intervener being a 'therapist' and the recipient being a 'client'. As argued in Chapter 2, the 'therapist' can be a politician who promises to improve our lives if we vote for him or her, a religious leader who promises that joining his or her faith will provide us with life after death, a psychotherapist who offers to solve our psychological problems, and a criminal who pretends to be a 'therapist'.

But not all 'therapists' are manipulators.

7.4 Participation in Communication as the Mechanism of Manipulation of Living Consciousness

Under *magical manipulation of living consciousness* we will understand the intervention based on the PC, which aims at relaxing the person's reason and reach the person's living consciousness. The magical manipulation should be distinguished from *ordinary manipulation*, which is based on coercion (for example, when a marriage partner manipulates his or her spouse using the power of muscles or law, or a criminal manipulates its victims using a threat or clever deception). And even magical manipulation of consciousness is not necessarily ill-intended.

As mentioned in Sect. 7.2, in modern industrial society the normal way of intervention is to address the person's 'observer' with a logically grounded offer. This is where rational communication differs from manipulation. For example, in today's medicine a doctor must rationally explain to a patient why taking certain drugs will bring a cure. In contrast, in traditional tribal societies the healing effect is achieved due to the patient's belief in a medicine man's magical powers [3]. However, the PC works in modern medicine as well, in the form of the *placebo effect*.

The placebo effect is achieved when the patient is given a capsule filled with pure water and told that the capsule contains medicine that will make the patient feel better. Sometimes after taking such a capsule, the patient really does get better. Surprisingly, an improvement in the psychological state can occur even if the doctor says that the capsule contains pure water, but the patient will still feel better (the so-called 'placebo without deception') [7]. This shows that although the patient's 'observer' understands that the ingested liquid cannot bring a relief, the doctor's message passes through the 'observer's critical thinking and gets to the person's 'experiencer', who believes in the magical power of the doctor's words. Although placebo does not work for serious cases, it has a clinically measurable effect in relieving pain and other negative subjective experiences.

Even in everyday communication we often experience the PC effect. For example, if someone told us that today we would have a car accident, our 'observer' would treat this message with scepticism, but the 'experiencer' would still have an unpleasant sensation than the prediction might come true. By the same token, unjustified condemnations or praises from strangers sometimes reach our 'person-experiencer' and influence our mood and emotions. Fortunately, under normal circumstances such effects are too weak to affect our behaviour.

The placebo effect suggests that the PC could work even when a suggestion doesn't have a direct reference to magic. To test this possibility, participants in the psychological experiment (students and university staff) were divided into two groups [8]. In the *magical condition* of the experiment, participants were asked to imagine their future life and then suggested that the experimenter casts a magic spell on them that would affect their future life by making this life easy (a good spell) or difficult (a bad spell). In the *non-magical condition*, participants were offered the same message, but this time the message was free from overt association with magic: the participants were told that if the experimenter decreased or increased the number of 1s on a computer screen (for example, turning 111111 into 111 or vice versa), then the number of difficult problems in the participant's future life would decrease (a desirable outcome) or increase (an undesirable outcome) proportionately. In both conditions, the participants reasoned rationally and argued that neither the magic spell nor the manipulation of the numbers on the screen would affect their lives. However, in practice, the non-magical suggestion worked the same way as the magical suggestion: about half of the participants allowed the good spell and the decrease in the number of units on the screen but almost all participants prohibited the bad spell and the increase in the number of units on the screen. These data confirmed the prediction that *ordinary suggestion*, like magical suggestion, is based on the PC: the unconscious belief that a verbal statement of an authoritative person can, contrary to logic, come true.

All and all, the experiments showed that both *magical and non-magical psychotherapies are based on the PC*. The first 'psychotherapists' in history were wizards and priests, and the concept of suggestion grew out of ancient magic, changing the form, but not the content.

7.5 Techniques of Magical Manipulation Based on the Participation in Communication

How do the 'therapists' manage to deprive reason of its ability to control person's actions and get through to the person's living consciousness? For this, special techniques are used, the purpose of which is to activate the PC. This is achieved by the 'therapists' pretending to be people with special abilities or social positions: civil servants, representatives of law enforcement, bank employees, successful businessmen, etc. (Condition 1 of the PC activation: the person's ***trust in the manipulator's special powers***). At the same time, the 'therapists' break through the person's critical thinking by creating conditions under which employing critical judgement for decision making involves high cost for the person. The breakthrough is achieved by addressing the person's personally important entities, values, and passions, for example, by saying that the person's close relative has had an accident and if the relative is not helped *now* he or she may die (Condition 2 of the PC activation: ***debilitating the person's critical thinking*** via pitting the high cost on doubting that the message is true). As a result, the person's internal objectified consciousness (the buffer zone) is temporarily incapacitated and the person follows the 'therapist's message in spite of being aware that the message is untrue or ill-intended. Here are some well-known examples of this kind of 'social engineering'.

- *Economy.* The so-called 'Ponzi scheme'—a financial organization pays interest to its investors from the new capital provided by new investors rather than from the profits earned by the organization [9]. Such a scheme, or a pyramid, collapses over time, as maintaining high interest rates requires an ever-increasing flow of funds from new investors. Investors are lured into the pyramid scheme because (1) They believe in the pyramid leader's unusual ability to make them rich quick, and (2) The cost of doubting such belief is high (loss of funds already invested and hopes of getting rich quick and easy).
- *Psychoanalysis.* In psychoanalytic therapy, the key to success is the so-called 'transference effect' when a patient comes to view an analyst as having unusual abilities and begins trusting the analyst. As one of the authors writes, "The patient's perception of the therapist and expectations from the therapist become unusually high and not adequate to reality" [10, p. 39]. This provides Condition 1 for the PC activation—the belief in the therapist's extraordinary abilities. Condition 2—the high price to pay for the lack of unconditional trust—is provided by the requirement that "patients must be strongly motivated to overcome their difficulties" [ibid., p. 40]. As a result, the patient can become the object of magical manipulation and take on faith any, even the most arbitrary, interpretation of his or her statements by the therapist.

- *Neuro-Linguistic Programming (NLP).* In psychotherapy known as NLP, the first act of therapy is to establish a relationship of trust, or rapport, between a therapist and a client. The therapist achieves this trust by attuning to the client and mirroring his or her verbal and non-verbal actions. For example, if the client has a kinaesthetic system of representing the world in his or her living consciousness, the therapist builds communications with the client based on kinaesthetic predicates, thus making the client feel that the therapist understands him or her and can help with his or her problems [11]. Having gained the clients' trust, the therapist creates Condition 1 for the PC activation: the clients' belief in that the therapist has the extraordinary ability to solve their problems. The high price for refusing to believe in the therapist's extraordinary abilities is the loss of the clients' hope to get rid of their problems. This creates Condition 2 of releasing the client's PC. The founders of NLP claim that NLP can treat a wide range of physical and mental ailments, such as learning disabilities, phobias, schizophrenia, multiple personality disorder, depression, and post-traumatic stress [12]. While some of these conditions (for example, phobias) can indeed be alleviated or cured based on the placebo effect, numerous test studies have not confirmed the claim of NLP to treat more serious diseases (for example, schizophrenia) and characterize NLP as pseudoscience or quasi-religion [13–15].

Psychoanalysis, NLP, and some other types of psychotherapy can be useful interventions in treating clinical conditions based on subjective feelings of patients (such as stress or phobias), since such conditions are based on the placebo effect. But if psychotherapy crosses a certain boundary (for example, denies psychiatric illness as requiring the use of drugs), it can be potentially harmful since it distracts the patient from treatment with conventional medicine. A lot more influential, and therefore harmful, can be magical manipulations of living consciousness of groups and masses.

7.6 Magical Manipulation of Living Consciousness of Groups and Masses

In all the cases listed below, both conditions of activating the PC were observed to make magical manipulation possible: (1) followers believed in the unusual abilities of their leaders, and (2) the spiritual, economic, or political cost of doubting such belief was high.

- *Religion*. In November 1978, the leader of American religious organization 'People's Temple' James Jones convinced 909 of his followers to commit suicide in Georgetown, Guyana.

In 1993, David Koresh, the leader of a religious sect based in Waco, Texas, convinced his followers to fight to the tragic end against FBI agents besieging their farm.

In March 1997, Marshall Applewhite, the leader of UFO religious cult based near San Diego, California, convinced 38 members of the organization to commit suicide in order to travel to an alien spacecraft following comet Hale-Bopp.

- *Politics*. In the course of 'perestoyka' most of Soviet population uncritically accepted the suggestions of their governing elite that privatization of public property will bring prosperity. In reality, the privatization was a way for the Communist party and the KGB upper crust to appropriate people's property for their own enrichment [16].

Strong anti-war opposition in the British Parliament proved unable to stop the 2003 invasion of Iraq. Despite the rational arguments against joining this military campaign, the Prime minister of that time conceded the Parliament's approval [17]. Similar protests against intervention in the civil war in Libya in March 2011 failed as well [18]. The drastic consequence of these decisions—a massive migration of displaced and impoverished people to Europe—is still shaking the EU today (see Chapter 8 for more on magical manipulation with minds in politics).

- *Terror*. Brainwashing techniques for training suicidal terrorists are based on the PC: the victims of such training are made to believe that (1) their actions are controlled by God, and (2) the cost of doubting such belief is high (loss of opportunity to become a martyr and go to heaven) [19, 20].

7.7 Protection Against the Magical Manipulation of Living Consciousness

7.7.1 *Rule 1. Watch Your Feelings*

Practice shows that warning potential victims of the danger of being magically manipulated doesn't work. It is necessary to explain to people that manipulators will make the first blow to their feelings and emotions, such as the fear of losing loved ones, the desire to become an object of care, or the thirst for quick riches.

If you feel that the 'therapist' has elicited in your fear or craving for something, you may become the object of magical manipulation.

7.7.2 *Rule 2. Remember that Magical Thinking Still Lives in Your Unconscious*

When a 'therapist' (government official, psychotherapist, commercial organization, political or religious leader) tries to gain your trust, remember that your trust can go beyond ordinary respect and develop into faith in the magical powers of the therapist.

To prevent this from happening, always pay attention to the methods by which the therapist tries to gain your trust. See if the therapist provides rational arguments about why he or she can be trusted or appeals directly to your feelings and emotions, claiming that he or she has unusual powers or abilities.

If the therapist emphasizes his or her extraordinary powers or abilities the nature of which you do not understand, it is possible that you became (or will become) the object of magical manipulation.

7.7.3 *Rule 3. Pay Attention to the Request for Unconditional Trust Without the Right of Criticism*

Usually, to establish a trusting relationship with a client, the respectful attitude of the client to the therapist is required, while the client is not prohibited from criticizing the therapist and asking questions. However, sometimes the therapist demands unconditional trust from the client without the right of criticism, claiming that otherwise the therapy will not work.

If the therapist demands from you to give up the right of asking questions, it is likely that you became (or will become) the object of magical manipulation.

7.7.4 Rule 4. Watch for Signs of Ambitious Superiority in the Therapist's Behaviour

If the therapist, consciously or unconsciously, seeks to magically manipulate your living consciousness, he or she may inadvertently betray such a desire, emphasizing his or her physical, mental, or moral superiority over you, or his or her important social status. If at any stage of therapy you feel that the therapist exalts himself or herself above you or emphasizes his or her special merits, it is possible that you became (or will become) the object of magical manipulation.

7.7.5 Rule 5. Monitor the Possibility of Interrupting Therapy at Any Time and Consult with a Third Party

During therapy, you should be given the opportunity to interrupt and exit the therapy without consequences for you. For example, you may ask for a break for a telephone consultation or interpose the therapy in order to think and get someone else's advice.

If the therapist prevents you from interrupting the therapy or makes the interruption conditional (for example, you lose a lot of money or the opportunity to be cured of a serious illness), it is likely that you became (or will become) an object magical manipulation.

7.7.6 Rule 6. Never Give in to Pressure

Sometimes during therapy, the therapist will ask you to do some kind of exercise that confuses you (for example, actions that violate generally accepted moral or sexual taboos). However, the therapist insists, arguing that doing so is necessary for success of the therapy.

If at any time during therapy you have an internal moral conflict with the demands of the therapist, it is possible that you became (or will become) the object of magical manipulation.

7.7.7 *Rule 7. Always Weigh Pros and Cons in Advance*

Before you sign a contract with a therapist or let him or her into your home, think carefully about whether this therapy is free from the possibility of magically manipulating you. Ask yourself the following questions:

Are the results promised to you by the therapist achievable?
Is the price of the therapy proportional to the promised results?
Does the therapist explain principles and techniques of the therapy to you?
Does the therapist seem an honest and decent person to you?
Have you heard anything positive about this therapy from people you trust?
If the answer to any of these questions is no, follow Rules 1–6.

7.8 Participation in Communication as a Base of Magical Manipulation: Conclusion

In this chapter, we considered the case of manipulation with living consciousness of individuals, groups, and societies based on personal contact with the victims. However, the PC mechanism can have a wider application. In the next chapters, we will analyse the PC in more detail, and see how this mechanism of magical manipulation of living consciousness works in wider impersonal domains, such as the domain of fake-reality, totalitarian media narratives, and the impersonal living consciousness of societies.

References

1 https://en.wikipedia.org/wiki/Reason
2 Subbotsky, E. (2010). *Magic and the mind. Mechanisms, functions, and development of magical thinking and behaviour*. Oxford University Press.
3 Lévy-Brühl, L. (1985/1926). *How natives think*. Princeton University Press.

4 Subbotsky, E. V. (1993). *The birth of personality: The development of independent and moral behaviour in pre-school children*. Prentice-Hall.
5 Subbotsky, E. (2005). The permanence of mental objects: Testing magical thinking on perceived and imaginary realities. *Developmental Psychology, 41*, 301–318.
6 Milgram, S. (1992). *The individual in a social world: Essays and experiments* (2nd ed.). McGraw Hill.
7 Kaptchuk, T. J., Friedlander, E., & Kelley, J. M., et al. (2010). Placebos without deception: A randomized controlled trial in irritable bowel syndrome. *PLoS One, 5*(12), e15591. https://doi.org/10.1371/journal.pone.0015591
8 Subbotsky, E. (2007). Children's and adult's reaction to magical and ordinary suggestion: Are suggestibility and magical thinking psychologically close relatives? *British Journal of Psychology, 98*, 547–574.
9 https://ru.wikipedia.org/wiki/Ponzi,_Charles
10 Arlow, J. A. (1989). Psychoanalysis. In R. J. Corsini & D. Wedding (Eds.), *Current psychotherapies* (4th ed., pp. 19–62). Peacock Publication.
11 Grinder, J., & Bandler, R. (1976). *The structure of magic II*. Penguin.
12 Bandler, R., Grinder, J., & Andreas, S. (Eds.). (1979). *Frogs into princes: Neuro linguistic programming*. Real People Press.
13 Sharpley, C. F. (1987). Research findings on neurolinguistic programming: Nonsupportive data or an untestable theory? *Journal of Counselling Psychology, 34*(1), 103–107. https://doi.org/10.1037/0022-0167.34.1.103
14 Witkowski, T. (2010). Thirty-five years of research on neuro-linguistic programming. NLP research data base. State of the art or pseudoscientific decoration? *Polish Psychological Bulletin, 41*(2). https://doi.org/10.2478/v10059-010-0008-0
15 Thyer, B. A., & Pignotti, M. G. (2015). *Science and pseudoscience in social work practice* (pp. 56–57, 165–167). Springer Publishing Company. ISBN 978-0-261-7769-8.
16 Subbotsky, E. V. (2013). Russia in a conjured dream. *Psychologist, 10*, 40–83. https://nbpublish.com/library_read_article.php?id=11353
17 https://www.theguardian.com/politics/2003/mar/18/iraq.iraq6
18 https://en.wikipedia.org/wiki/Protests_against_the_2011_military_intervention_in_Libya
19 Ivanov, U. G. (2001). *Kamikadze: Piloty – smertniki. Yaponskoe Samopozhertvovanie vo vremja voiny na Tikhom Okeane*. Pusich. https://libking.ru/books/prose-/prose-military/206718-yuriy-ivanov-kamikadze-piloty-smertniki.html
20 Atran, S. (2003). Genesis of suicidal terrorism. *Science, 2999*, 1534–2153.

CHAPTER 8

Fake-Reality: Cheating on Living Consciousness on a Mass Scale

8.1 Magical Participation with External Messages as a Source of Fake-Reality

In the previous chapter we discussed the mechanism of the PC when the source of the communicative exchange is another person, who we designated as a 'therapist'. However, in the everyday life we are targeted by a flood of stimuli which are impersonal: adverts on TV and walls of the tube stations, news in newspapers, paintings and sculptures in museums, food in restaurants, aromas in the perfume shops, and most of these stimuli are messages addressed to us and asking us to watch, read, admire, believe, eat, smell, buy, and consume in any other way. Some of these messages we accept and consume on a regular basis, others only once, still others leave us indifferent. So, what ability of our living consciousness makes us treat these messages differentially? This ability is *participation with external reality (PER)*. Let us consider this ability in more detail.

Earlier in this book we described the magical laws of sympathy and contagion, when certain objects that are physically separate but related to each other (for example, a person and his or her photograph or a clump of hair) are magically connected forever and at any distance. However, in living consciousness operates one more magical law that is considerably less studied—the law of participation. According to this law two unrelated objects that don't resemble each other and have never been in contact with each other could, nevertheless, share a common essence or soul. For

E. Subbotsky, *The Magic of Living Consciousness*,
https://doi.org/10.1007/978-3-031-54708-9_8

example, Bororo people of Central Brazil mystically identified themselves with a certain species of parrots, which they viewed as their totem. It was not that Bororo did not see a difference between a person's physical appearance and that of the parrot; yet Bororo maintained that a spirit of the parrot lives in a person, and so in each Bororo there is a part of the parrot, and vice versa [1].

In a similar way, through participation, external symbolic messages may become part of our living consciousness. According to Freud and some modern theorists, our unconscious living consciousness operates not through concepts, but through images [2]. But images are also the language of the conscious living consciousness. When external messages are shaped in the form of images, they can reach our living consciousness via the participatory-magical route (see Fig. 6.4). But not only images could penetrate our living consciousness bypassing the buffer zone of critical thinking, spoken and written words can do this as well.

In the previous chapter we assumed that a possible way to account for the origins of human language is to view early language as a form of a magic spell that inherently contained suggestive power. In the beginning, early humans began to attribute humans and objects with having invisible spirits inside, and thus invented the way of thinking and talking about entities that are not there by addressing the spirits, as if the addressees were in fact present. As the animistic beliefs faded, the spirits were converted into *meanings*—the cognitive constructs that were no longer associated with the invisible living entities hidden in objects. But even today most languages bear the traces of their animistic origins. When we say, 'the sun is rising' or 'the engine is working', we don't really mean that the sun and the engine are living entities capable of active goal-directed behaviours, yet we speak as if they were. The anthropomorphic nature of language indicates that once a spoken word was a magic spell addressed to an invisible sentient entity—the spirit, and the message was supposed not to be critically evaluated but obeyed. And when the message was pronounced by God, it was identical with reality: 'And God said, "Let the waters under the sky be gathered into one place, so that the dry land may appear." And it was so' (Genesis 1:9). When modern languages turned into a system of rational meanings, to some extent they retained their suggestive participatory power. Psychological experiments reviewed in Chapter 7 revealed the presence of the PC in our living consciousness: even if consciously we understand that a suggestive message is wrong, unintentionally our internal Self might still identify this message with reality and obey commands that this message implicitly contains.

All and all, both historically and psychologically, the PC could be the 'raw material' from which the rational communication was shaped, by eliminating the word/reality identification that a semantic message inherently contains [3]. However, this elimination is never complete. That is why commercial advertising occasionally gets through to our living consciousness and makes us comply even if complying is rationally unprofitable to us. The persisting tendency of our internal Self to unintentionally identify itself with external symbolic messages via participation—the PER—was the psychological ground for the emergence of a special type of media reality—*fake-reality*.

8.2 Fake-Reality

Almost every day we read and hear about fake news. The number of studies on why people distort information and believe in it has already exceeded hundreds [4, 5]. But just look back at our daily lives and it becomes clear that fake news is nothing but a tip of the iceberg on which we float in the ocean of life. The name of this iceberg is 'fake-reality', or simply fakes.

Fake-reality is a reality that only pretends to be real. Here are some definitions of fake, taken at random from dictionaries. Fake is 'anything made to appear otherwise than it actually is' [6], 'artificial fur or fake painting ...is something that was made to look valuable or genuine, often to deceive people' [7], 'an object that is made to look real or valuable in order to deceive people' [8]. Fake is not the same as fiction. Conceptually, fiction is a product of human imagination which doesn't pretend to be reality. For example, novels by Fyodor Dostoyevsky and paintings by Salvador Dali are fictions yet not fakes, but forgeries of paintings by great artists passed off as the originals are fakes. The Crimea's sparkling wine is a fiction, but the Crimea's champagne is a fake, as it pretends to be a sparkling wine that is produced exclusively in the Champagne region of France from established grape varieties. And although both fictions and fakes are types of invented reality, they differ in that fakes pretend to be someone or something they are not, and fictions don't. Likewise, not every copy is a fake. For example, a copy of an antique sculpture labelled 'a copy' is a copy, and the same copy labelled 'the original' is a fake. In relation to poor, low-quality fakes, which are the majority, the Latin term 'simulacrum' is often used in philosophy.

In the everyday life fakes are most often found in the domain of morality: intended lying is an attempt to present something non-existent and pass it off as reality. Another domain that provides fertile ground for fake-reality is health protection. In almost every newspaper, we see articles about new miracle devices and services that, in the shortest time and without effort on our part, will help us to lose weight, get rid of diseases, cleanse the intestines of parasites, and defeat autumn depression. On the Internet, pictures of hands, feet, and other parts of the body flicker, the treatment of which requires immediate purchases online. The natural suspicions of potential customers that these devices and services have not been verified and may be fakes are neutralized by the attached (faked) 'customer reviews' and 'questionnaires'.

In the domain of industry, fakes of well-known brands of consumer goods abound. In the domain of service, it became difficult to determine whether a particular firm is a genuinely solid institution or a hastily formed fake, providing services in the field of electricity, household appliances, or plumbing, the effect of which equals to zero and the cost is substantial. In the domain of culture, the situation is no better: authors 'borrow' ideas from others without referring to the originals, works of art are forged, invented and unverified data are reported in scientific journals, museums display sandbags, tin cans, and urinals under the guise of works of art. In the domain of cosmetics and other sensual experiences, fakes have reigned since the time of Adam. Women paint their lips and nails to appear prettier than they really are, perfumes and fragrances create fake aromas, dyes make food more attractive to the eye, and flavourings and spices deceive our taste.

But the real fake revelry is observed in the domain of information. Historical events in textbooks can be replaced by fakes in favour of the current political agenda. In the media, news reports almost always magically match the narratives of competing parties which are opposed to each other. Phone and other types of fraud, in which criminals pretend to be law enforcement officers or representatives of other respected professions, have become catastrophic, especially against the elderly. Email is flooded with fake messages and ads, and virtual reality allows you to 'attach' head of one person to the body of another without waiting for medical scientists to learn how to transplant a real head.

Finally, in the domain of geopolitics, fakes can occur on a scale of entire enclaves if they are forcibly synthesized into states without a federal political structure.

Summarizing, one can say that people in modern industrialized countries live in fake-reality. Questions arise, such as 1) what are the psychological mechanisms for the emergence of fakes and 2) has reality always been like this, and if not, why is reality, which in the earlier epochs used to rely on truth, albeit a bitter one, nowadays is becoming increasingly fake?

In order to answer these questions, it is necessary to make excursions into individual and cultural development of living consciousness.

8.3 Reality One Can Play with

Developmental studies have shown that from about the age of about 3 years children are able to give a plausible oral report of events that had happened to them in the last few minutes [9, 10]. Let's call this report 'a master record' or simply 'tape' (abbreviated as 'T'). Unlike an ordinary tape recording, the 'T' is an active process, but it has an involuntary and compulsory character. Up to a certain age, having living consciousness based on the 'T' (the 'T- living consciousness') is enough for a child to communicate with other people, because the child's personal interests usually coincide with those of close adults. The situation changes at the age of about 4 years, when there appear situations when children's personal interests diverge from those of others: The child discovers that if he or she presents his or her 'T' to close adults they won't approve of it.

In our analysis of the 'divergent interests situations' we omit the trivial cases when the child is too young to be able to control his or her actions. For example, the child's mother prepared a cake intended for guests and left the cake on the table unattended, having warned the child that the cake should be kept untouched until the guests arrive. Clearly, the 2-year-old child will be unable to resist the impulse, and most adults know this. As children grow older, the adults begin to expect more obedience from their children and are more likely to exhibit their displeasure when the children disobey prohibitions.

The awareness that doing prohibited actions elicits negative responses in close adults becomes a trigger that makes older children's living consciousness more complex: the children develop the ability to view their own actions through the eyes of other people, thereby anticipating the possible negative reactions of adults to their transgressions. In the attempt to find the way out of this conflict with close adults, the children make a discovery that to satisfy their desires and avoid punishment they must create a special story intended for the outside world—the Fake Tape story (the 'FT-story')—for example, that it was their dog who bit off a piece from the cake. Since creation of the 'T' is the default process, the 'T' cannot be edited. With the 'FT-story' the situation is different. Here the child deliberately changes and edits the events, adapting the 'FT-story' to fit his or her needs. As a result, the children develop the ability to create fake-reality in the form of deception. Having mastered the ability of creating the 'FT-stories', children develop the new type of living consciousness—the 'FT-living consciousness'.

Psychological studies of moral behaviour showed that, given the opportunity, about 80% of 3- to 7-years-old children prefer to deceive another person by creating the 'FT – story' [11], and in adults the proportion of people inclined to use fakes in similar circumstances is about the same [12–14].

All and all, most people of all ages, except for the very young, are psychologically disposed to use fakes for personal gain if the gain exceeds the harm from punishment. The ability of people's living consciousness to invent the FT-stories is the psychological basis of the production of fake-reality.

However, the mere ability to create fake-reality for selfish purposes is not enough to explain why the fake-reality became so widespread in the twentieth century. Indeed, psychologically modern people are not so different from their ancestors of a few past centuries. Perhaps, the intensification of fake-reality in the modern time can be explained by some changes in culture.

8.4 Fake-Reality in a Historical Perspective

As argued in Chapter 6, before the twentieth century European culture, by and large, maintained the beliefs in that living consciousness reflects reality and human behaviour is monitored by divine supervision. These beliefs meant that creation of fakes, such as a false accusation, is a violation of the will of God and deserves punishment from both God and people (e.g., 'Do not bear false witness', Mt. 19:18). Gradually, the association of fakes with religious prohibitions began to crumble, but the creation of fakes was still viewed as a morally unacceptable distortion of reality.

In the play 'The Dowry' by the Russian playwriter Alexander Ostrovsky (1823–1886), the merchant refuses the possibility to satisfy his passionate desire because his 'merchant's word' given by him to another merchant prohibited this. This example suggests that although people have always been psychologically inclined to use fakes for personal gain, the common view in the nineteenth century was that people must obey the norms of law and morality.

However, in the middle of the twentieth century the philosophy of subjectivism, which until then has been circulating among a thin layer of educated intellectuals, began to penetrate the deep layers of culture, leading to a decline of the faith in 'realism' and in the divine retribution for the sin of lying. In mass culture, this trend gave rise to simulacra—images which penetrated all spheres of public life. According to the French philosopher Jean Baudrillard (1929–2007), simulacra creates a hyperreality that duplicates and replaces genuine reality. In everyday life today examples of simulacra are financial pyramids, fake money, false eyelashes and nails, flavours, and food imitators. As a result of the belief in that living consciousness not simply reflects reality but partially creates it, fake-reality turned from a violation of the laws established by God into a legitimate part of the daily life. It is not surprising then that what in the past centuries was called lies is now called disinformation, social intelligence, 'social engineering', or the art of diplomacy.

But not all fakes are harmful or useless in life.

8.5 Fakes Good and Bad

In real life, we are sometimes forced to lie and pretend. I recently watched a BBC television programme. The interviewer, a prominent British journalist, asked difficult political questions to a senior NATO representative. The face of the interviewee expressed a sense of tension and caution. The interviewer asked questions requiring 'yes' or 'no' answers, and the interviewee tried his best to avoid direct answers. As soon as the interview ended the interviewee's face relaxed, throwing off the mask it had been wearing 'for the public'. It was obvious that what the interviewee really thought and what he presented publicly were two different stories. Without fake stories created for public consumption, social institutions such as politics, diplomacy, civility, and etiquette would collapse. And remember how often we have to tell our loved ones not what we really think or feel but what is needed in order not to disturb peace in the family.

Indeed, some symbolic actions, such as gifts, touches, kind words, and imitation of feelings, can do miracles. In everyday life, each person can be a 'magician', giving gifts and saying kind words to other people, not because he or she has sincere feelings towards those people, but in order to preserve or existentially renew fragile relationships. In personal relationships, the natural mechanisms that hold relationships together, such as sexual and physical attraction, can weaken over time, while objective factors, such as maintaining a family and raising children, require that relationships continue. In such circumstances, creating the illusion of feelings through saying kind words, giving presents, and observing love rituals can help replace natural mechanisms and enhance the existential status of that fragile and delicate phenomenon of magical participation that we call love. In the same way, civility and diplomacy in international relations are intended to replace the absence or degradation of the natural economic and political interests that usually bind people and countries together. When a person becomes a target of a message that mimics love, politeness, or diplomacy, it is beneficial for him or her to pretend 'believing' in sincerity of the message. For example, if in the middle of a transatlantic flight a tired stewardess brings you a glass of champagne at your call, you might realize that the feelings the stewardess experiences towards you are not exactly love and sympathy, but both you and the smiling stewardess prefer to ignore true feelings and take fake feelings for reality, as this helps you to feel comfortable and the stewardess to keep her workplace.

Less unequivocally useful, but not harmful, are fakes in the field of imitation of sensory impressions. Cosmetics, fragrances, deodorants, flavour enhancers, and other sensual fakes make our sensations more intense and pleasurable, albeit not for long. For example, skilfully dyed hair, which seemed so magical during the romantic period of courtship, can evoke completely different feelings when, after the conclusion of a legal marriage, the hair's true colour is revealed.

However, much of the fake-reality is harmful and dangerous. In Chapter 7 we analysed the rampant scam that uses fake 'care' for people to take possession of their property. A variety of this kind of fraud is fake organizations in the service sector or the production of fake drinks and products. Let's remember how many people were poisoned by counterfeit vodka during the campaign against alcoholism in the USSR. A version of fake-reality is corruption—the misappropriation by civil servants of communal property under the pretext of complying with government orders.

Not only an individual, but also large groups of people can become puppets of fake-reality. One form of fake-reality is fake social narratives. Like genes and memes, social narratives are myths, hyper-individual forces that feed on beliefs and consciousness of individuals. Not all social narratives are fakes, but those that elevate an abstract idea to the level of universal truth, such as the myth of the victory of communism all over the world or the myth of the superiority of the Aryan race, are.

Political fake narratives are joined by economic, religious, and scientific ones. Financial pyramids and bloggers empty the wallets of the gullible, deceptively promising them quick enrichment through the use of the elemental forces of a market, and the founders of religious sects turn their followers into puppets, presenting themselves as mentors blessed by God. Even science data can become the basis of fake social narratives, for example, in order to force poor countries to stop using certain energy sources for improving the climate or encourage people to purchase insufficiently tested medicines.

So, what should a person do when he or she is forced to live in fake-reality?

8.6 Living with Fake-Reality

A recent newspaper issue described the case when scammers had managed to persuade a professor from a reputable university to join them in a 'catching scammers' exercise, as a result of which the professor lost his money, his summer house, and his apartment. But why a cultured and educated person is helpless in the face of much less educated deceivers? Perhaps, for a person living in an environment where most people obey the laws of honour and morality it is not easy to accept the fact that along with real people fake people walk around. Fake people don't look like criminals, don't use foul language, don't abuse drugs, and are decently dressed, but they are free from compassion, rules of honour, and moral self-esteem. Unfortunately, the scammers know well psychology of normal people and skilfully use this knowledge to their benefit.

In other words, we are vulnerable to fake reality because at schools we were not taught a subject that should have been introduced in the middle of the twentieth century—'fakeology', or the *science of fakes*. This science should have taught us that people's living consciousness is inherently vulnerable to participation with symbolic messages—the PC and the PER, and that this property of living consciousness is widely used in poetry, politics, and commercial advertising. Unfortunately, this property can also be abused by fraudsters. Today fake-reality is an inevitable and even a necessary part of our daily life, and from our early years we need to train ourselves in how to separate useful fakes from harmful ones, using the former and resisting the latter.

Thus, one shouldn't resist fake politeness and diplomacy in personal and public relations under the pretext of 'fighting pretence', since the exposure of negative feelings in a competitive world could lead to escalation of conflicts. Conversely, the ability to create positive fakes in the form of civility and diplomacy increases your competitiveness. At the same time, it is necessary not to cross the red-line, so that the fake politeness and goodwill are not perceived by others as a desire to 'suck up' to the authorities and sacrifice one's own dignity for the sake of momentary gain.

As already mentioned, in interpersonal and intimate relationships fakes in the form of symbolic gestures, inexpensive gifts, nice words, and invitations to the theatre or café should not be neglected—even if these actions are not caused by sincere emotions. Such gestures require little effort, but the result can be the maintenance of meaningful and rewarding personal relationships. If, from your point of view, people significant to you behave incorrectly or cause you unpleasant feelings, it is more useful to avoid personal contact under a fake pretext than to give away your true feelings. Finally, if you disagree with the opinion of a speaker at a public talk or an author of an article and feel angry towards him or her, you should not express your negative feeling in the form of offensive comments but rather express disagreement in the form of a positive fake (e.g., sympathy for the author's unsuccessful attempt to cope with the difficulties of the problem under discussion). It must be understood that insulting an opponent is also a fake. Although your offensive comments may seem to you a struggle for truth, these comments mask the hidden sense of your intellectual inferiority and in a public perception might work against you.

On a larger social scale, it should be remembered that people tend to join social narratives not only due to circumstances or for personal gain, but also voluntarily, in order to avoid the feelings of loneliness and insecurity. Sometimes it is difficult for a person to decide which of the conflicting social narratives is real and which is fake. Even if a moral law is on the side of one of the conflicting narratives and the opposite one is a fake, this difference may be hidden from view and the fake social narrative seduces many via manipulating the data of science and religion.

For example, the fight against the fake narrative of Nazism in the middle of the twentieth century was long and difficult and costed millions of lives.

It is also possible that both conflicting social narratives are fake. In such a situation, taking side of one of the conflicting parties indirectly contributes to escalation of the conflict and multiplication of its victims (see Chapter 9 for the elaboration on that). Consequently, if you are not involved in one of the conflicting social narratives by force, the optimal position in relation to social narratives with unclear 'fake status' is the position of non-alignment and a call for solving the conflict through compromise and negotiations. One must bear in mind that even the uncompromising victory of one of the conflicting social narratives is never final and leads to a temporary attenuation of the conflict. The end of the conflict comes only when the tectonic forces of history sink both conflicting narratives into the river called Lethe and replace them with new ones. For example, the victory in the war against the fake narrative of Nazism in 1945 did not destroy Nazism, and the battle of narratives for the French throne between England and France, although it lasted for more than a hundred years, ended in the fifteenth century [15].

8.7 Participation and Fake-Reality: Summary and Conclusion

In this chapter, we considered how the mechanism of magical participation with external reality (PER) works regarding impersonal messages that surround a person in the modern world. The psychological mechanism of the PER is the same as that of the PC and includes the 'trigger' and the 'gunpowder'. In the PER the trigger is a piece of impersonal symbolic objectified consciousness in the form of advertisements, news, and other printed or displayed messages, and the gunpowder is the unconscious residue of the ancient belief in the identity between a message and reality. Like in the PC, in the PER the degree of a person's participation with the message depends on the authoritative status of the message. Thus, we are touched by good poetry but indifferent towards poor one. We trust news that come from reputable sources and not from unknown sources that abound on the web. To overcome the barrier of our critical thinking, fraudsters create messages that mimic messages from respectable sources. These massages, or fakes, can reach our living consciousness bypassing the buffer zone of the internal objectified consciousness, or breaking through this zone. The mass production of fakes is the post-modern phenomenon that results from the crumble of the beliefs in realism and divine retribution.

With the boost of mass media via TV and the Internet in the second half of the twentieth century the fake social narratives that have been lurking for centuries in the living consciousness of individuals gave rise to a new phenomenon—the 'totalitarian media narrative'. In the next chapter, this phenomenon will be analysed in detail.

References

1 Lévy-Brühl, L. (1985). *How natives think*. Princeton University Press.
2 Mamardashvili, M., Pyatigorskii, A. (2009). *Simvol i soznanie: Metafizicheskie rassuzhdeniya o soznanii, simvolike i yazyke*. Progress-Traditsiya.
3 Subbotsky, E. (2023). *Examining the psychological foundations of science and morality: Explaining the inexplicable*. Routledge.
4 Pennycook, G., & Rand, D. G. (2021). The psychology of fake news. *Trends in Cognitive Sciences, 25*(5), 388–402. https://www.cell.com/action/showPdf?pii=S1364-6613%2821%2900051-6
5 Ecker, U. K. H., Lewandowsky, S., Cook, J., Schmid, P., Fazio, L. K., Brashier, N., Kendeou, P., Vraga, E. K., & Amazeen, M. A. (2022). The psychological drivers of misinformation belief and its resistance to correction. *Nature Reviews Psychology, 1*, 13–29.
6 https://www.dictionary.com/browse/fake
7 https://www.collinsdictionary.com/dictionary/english/fake
8 https://dictionary.cambridge.org/dictionary/english/fake
9 Belli, R. F., Schuman, H., & Jackson, B. (1997). Autobiographical misremembering: John Dean is not alone. *Applied Cognitive Psychology, 11*, 187–209.
10 Ceci, S. J. (1994). Repeatedly thinking about a non-event: Source misattribution among preschoolers. *Consciousness & Cognition: An International Journal, 3*, 388–407.
11 Subbotsky, E. (1993). *The birth of personality. The development of independent and moral behaviour in preschool children*. Harvester Weatsheaf.
12 Batson, C. D., & Thompson, E. R. (2001). Why don't moral people act morally? Motivational considerations. *Current Directions in Psychological Science, 10*(2), 54–57. http://www.jstor.org/stable/20182695
13 Batson, C. D., Thompson, E. R., Sueferling, G., Whitney, H., & Strongman, J. A. (1999). Moral hypocrisy: Appearing moral to oneself without being so. *Journal of Personality & Social Psychology, 77*(3), 525–537.

14 Bandura, A. (2002). Selective moral disengagement in the exercise of moral agency. *Journal of Moral Education, 31*, 101–119.
15 https://www.britannica.com/event/Hundred-Years-War/From-the-Treaty-of-Bretigny-to-the-accession-of-Henry-V-1360-1413

CHAPTER 9

Totalitarian Media Narratives: Living Consciousness in Captivity

9.1 The Belief in the Divine Origins of Political Power

If in the known Biblical story Jesus did not raise Lazarus from the dead but instead cured the badly ill Lazarus, would reducing the miracle to a class of rare yet possible events diminish peoples' unconditional respect to Jesus's divine origins? Why did gods need to be mighty wizards and not just magnified copies of ordinary people? Who needed gods to have magical powers—gods or people? It looks like people needed this more than gods. First, the people needed this in order to explain the unexplainable—the origins of the world and the human soul. But most important, people's leaders needed this in order to sanction their political and spiritual power. For instance, early Egyptians believed that the pharaoh was a son of God. The authority of modern spiritual leaders (e.g., the Pope) is believed to be sanctioned by God as well. It is possible to assume that the ancients' belief that political leaders are descendants of gods today transformed in the unconscious belief of modern people in the ***divine origins of political power*** *(BDOPP).*

In the everyday life today BDOPP could develop into unconditional trust in civil authorities, thus giving rise to the phenomenon of PC (see Chapter 7), but most prominently BDOPP reveals itself in people's trust in political leaders. A historically recent example of the work of BDOPP was the collapse of the USSR, when ideologists of 'perestroika' skilfully

E. Subbotsky, *The Magic of Living Consciousness*,
https://doi.org/10.1007/978-3-031-54708-9_9

used the Russian people's inclination towards BDOPP to privatize the public property. Unlike Stalin, who was adored by Russian people, the president of Russia Boris Yeltsin was not viewed as a living god, his human weaknesses were widely known and openly ridiculed by the population. And yet Yeltsin did what seemed to be impossible—cancelled the Soviet Union, put a ban on the Communist Party of the USSR, and transferred the communal property into private hands. Like Russian tsars, who passed their power to their heirs, Yeltsin passed his 'magic rod' of power to Vladimir Putin. That is why Putin—a mid-level officer of the KGB, unknown to the people and not a leader of a political party—suddenly became a victor of the presidential campaigns: Succumbing to their BDOPP, the majority of Russian people voted for Putin voluntarily, when the holder of the 'divine blessing'—Boris Yeltsin—said they should do so.

In a similar vein, many Western people today followed their leaders' appeals to initiate local wars and change regimes in independent countries, even though in private many of these people understood moral wrongness and futility of such enterprises and were sorry for the loss of lives of their soldiers. This apparently illogical behaviour of educated people should not be surprising if we accept that the peoples' BDOPP usually overrides rational considerations.

9.2 Democracy and Authoritarianism

The two main types of relations between the state and society—democracy and authoritarianism—are themselves heterogeneous. There is the democracy of Ancient Greece, from which slaves and women were excluded, and the democracy of the modern West. There is the authoritarianism of fear, as in the days of the Horde yoke in Russia or in countries occupied by Nazi Germany, and there is the authoritarianism of faith, when the people believe in BDOPP. And this is where psychology comes into play, in the form of the ***mental type*** of an average individual. The mental type is a tendency of an individual's living consciousness to succumb to BDOPP. Democracy arises in ethnic groups, in which personal experience of the individual's living consciousness prevails over the individual's BDOPP (mental type of an ***individualist***), while authoritarian societies mainly consist of individuals with living consciousness in which BDOPP has an upper hand over personal experience (mental type of a ***paternalist***). The authoritarianism of faith reveals itself in the phenomenon of PC, when people voluntarily submit to an order of an authority even if they consciously disapprove of the order and disobedience does not threaten them with reprisals. In modern Russia, for various reasons, a mental type of a paternalist prevails in the population, with the implicit belief that the higher political power is sanctioned by God (see [1] for the elaboration on that). After the October Revolution of 1917 religion was largely suppressed but BDOPP didn't disappear and was converted into the belief in a sacred right of the Communist Party and the classics of Marxism to determine the fate of people. Of course, there have always been people who thought differently—the intelligentsia, the open-minded ordinary people, but they did not make the state policy.

The essence of democracy lies in balancing the forces fighting within the living consciousness of the people. The democratic structure of society is not at the top of the ascending ladder of social progress, rather, it is a 'nation specific' way of linking people together. Speaking metaphorically, democracy is the point of the fragile balance between authoritarianism and anarchy. Even in societies where democracy is established it isn't forever. For instance, the Athenian democracy in Ancient Greece existed for about 300 years and gradually degenerated into oligarchic rule and authoritarianism. Similarly, the democratic institutions of ancient Rome descended into tyranny with the advent of the power of Caesars. Despite democracy literally means 'rule of the majority' [2]', it is the elite that has the upper hand in governing a democratic state, and democracy isn't a guarantee against criminality and corruption. The famous Russian philosopher Nikolai Berdyaev wrote, "The state exists not to turn earthly life into heaven, but to prevent it from finally turning into hell" [3]. Keeping a society from turning into hell could be done by both democracy and authoritarianism, depending on the mental type of individuals that prevails in the society.

9.3 The Totalitarian Media Narrative as a Phenomenon of the Society's Living Consciousness

With the boost of mass media as a result of the emergence of TV and the Internet, a new form of authoritarianism appears—the totalitarian media narrative (TMN). Generally, a TMN is a variant of a social myth, or a belief in what is happening in the world. A TMN can concentrate on natural processes (e.g., the belief that global warming is caused by the impact of civilization rather than by natural cycles of the planet), the causes of global epidemics (e.g., the belief in the artificial origin of COVID-19), the causes of social phenomena (e.g., the belief in the superiority of a particular race or nation over others), and processes in the field of education (e.g., the belief that studying classic languages is a waste of time and effort).

Since ancient times it has been obvious that the argument between conflicting social narratives is pointless. In the fifth century BC, the Greek sophist Protagoras is known to have said "of anything one can equally (ex aequo) argue from one side as well as from another (in utramque partem)" [4, p. 321]. The reason why the argument between two social narratives is unproductive is that the social narrative is rooted not in objectified consciousness (i.e., logical thinking) but in the living consciousness of the nation and is based on feelings and faith. Since social narrative resists logical and scientific arguments of the opposite social narrative, via undermining credibility of such arguments by means of real or far-fetched counterarguments, the institute of diplomacy fails. In addition to *insensitivity to the arguments of logic*, the social narrative also has such properties as expansiveness, inertia, selective absorptivity, stickiness, and infectiousness. *Expansiveness* is expressed in the fact that, starting in one domain (e.g., in politics), the social narrative spreads to economy (e.g., sanctions), culture (e.g., a prohibition on the display of certain works of culture), sports (e.g., a ban or infringement on the rights to participate in competitions), morality (e.g., a violation of contractual obligations and the creation of fakes), and even linguistics (e.g., an exclusion of images that might symbolize the competing social narrative). The *inertia* is that the social narrative emerges gradually, persists for a long time, and disappears or weakens slowly. *Selective absorptivity* is that the social narrative, like a dry sponge, absorbs all the data that confirm it, without requiring proof of these data and regardless of whether they are true or false (e.g., fakes), and rejects data that contradict it, even if these data are confirmed by independent experts. The *stickiness* of the social narrative is that if a person adopted a certain social narrative, it is hard for the person to refuse it even if the person wants to. Finally, the narrative tends to infect the opposite side with the desire to generate a counter narrative, with the same properties but in the opposite direction (*infectiousness*). As a result, the more confirming, albeit false, information a narrative produces or absorbs, the more successful it is in its impact on the mass living consciousness. Since fighting a social narrative by referring to logic and morality does not lead to success, the social narrative can only be neutralized by a counter narrative of equal saturation with supporting information, even if this information is a fake.

A person joins a certain social narrative for various reasons—to get rid of the feeling of loneliness, to overcome the state of uncertainty about what is happening in the world, or under the pressure from society. A totalitarian media narrative (TMN) is a special kind of social narrative which is spread by mass media. Like other social narratives, the TMN is rooted in faith and conditioned by a system of cultural and historical factors that go far beyond psychology.

For example, the current TMN of the West in relation to Russia presents Russia as something huge, intuitively talented, powerful, but politically backward and dangerous, and the current TMN of Russia in relation to the West presents the West as something culturally sophisticated, historically ancient, rich, but stagnating, tolerant towards violations of traditional Christian values and looking towards Russia with apprehension and desire for the weakening or even disappearance of Russia (of course, not everyone will agree with this description of the 'Russia versus the West' TMNs, but I leave to readers to make their own adjustments). The TMN is perpendicular to a country's political system and can arise in countries with authoritarian or democratic political structures alike. Importantly, the TMN is connected to the BDOPP; when this book is being written (the end of 2023), behind the TMN of the West the current President of the United States stands, and behind the Russian TMN stands the current President of Russian Federation.

The question arises what causes the emergence of conflicting TMNs? The cause is rooted in psychology. Generally, social narratives arise when FORM is violated. Any social narrative is immersed in specific events and is irrefutable by other narratives since each specific event can be interpreted in different ways. Only FORM stands above empirical events. For example, the biblical 'Thou shalt not kill' is a pure FORM. It does not specify who and under what circumstances should not kill, it says *no one* and *under no circumstances*. As soon as FORM is violated, the 'accuser and defender narratives' emerge. Since they are irreconcilable, a judge standing over them is needed. But in the conflict between sovereign nations there is no judge (see Chapter 10 for the elaboration on that).

As an example of the current and bloody conflict of social narratives is the conflict between Russia and Ukraine. Since this local conflict grew into the global conflict between Russia and the united West, which has been smouldering for centuries, let us consider the psychological origins of the local conflict first.

9.4 Russia and Ukraine: The Psychological Origins of the Conflict

9.4.1 *What FORM Was Violated in This Conflict?*

This FORM was formulated by the German philosopher Immanuel Kant in his categorical imperative "Act only according to that maxim whereby you can at the same time will that it should become a universal law" [5, p. 30]. Simply put, do things in a way that is beneficial not only to you, but to the whole of humanity. This FORM was violated on the maidan in Kyiv in February 2014, when a peaceful protest against the legitimate government of Ukraine escalated into the violent overthrow of this government. According to the FORM, people must respect the government they had democratically elected or replace it through peaceful protests and re-elections. Violation of the FORM happens when the peaceful protests develop into a revolution, because in the projection to the whole of humanity it leads to endless revolutions and chaos. Following FORM is necessary when we cannot predict the consequences of violating the FORM, and therefore following FORM saves us from ignorance of the consequences of its violation. For example, the people who overthrew the legitimate government in Kyiv in 2014 did not know the consequences of their action, could not predict that hundreds of thousands of Ukrainians would die in the war with Russia, and millions would become refugees. If that people followed the FORM, these drastic consequences would not happen. Despite tensions between Russia and Ukraine over the status of the city of Sevastopol where the Russian Black Sea Fleet was based, Russia and Ukraine coexisted on the basis of respect for each other's territorial integrity, millions of Ukrainians used to come to the Russian Federation as guest-workers, Russia gave Ukraine preferences in trade and supplied Ukraine with cheap oil and gas. The revolution of 2014 in Kyiv violated the FORM, and the two competing social narratives emerged: the social narrative of the new government of Ukraine (i.e., that the self-proclaimed Lugansk and Donetsk People's Republics are separatists supported by Russia) and the social narrative of Donetsk and Lugansk, which did not accept the violation of the FORM (i.e., that the new government of Ukraine is an unconstitutional usurper of power, supported by the US). This brings us back to psychology, since the social narrative is powered by the psychic energy not of the rational mind, but of the living consciousness in the form of faith. Let us consider the emergence of these core social narratives in more detail.

After the collapse of the USSR, parliaments appeared in both countries, and outwardly democracy was established, but the unconscious belief in the divine origins of political power (i.e., the BDOPP) did not cease to operate. In Russia, the effect of the BDOPP manifested itself in the centralization of power and transformation of the parliament into a consultative body under the president. It is sad to admit it, but the experience of the 2090s showed that the decentralization of power in Russia led not to democracy, but to activation of various organized criminal groups, rampant crime in the country, and the dangerous tendency towards further disintegration of the RF; so, in order to avoid the danger of collapse of the country and bloody civil wars, aggravated by the presence of nuclear weapon, the centralization of power, at least at that moment of Russian history, turned out to be an objective necessity. In Ukraine though, the situation developed according to a different scenario.

The people are always heterogeneous, consist of groups with different interests, and in order for these groups not to destroy each other, parties, parliaments, legislative and judicial powers arise in democratic countries, and all of them must be in a dynamic balance to each other. But such a balance implies responsibility—the subordination of groups to a legitimately elected government. When one of the groups refuses to take the responsibility and overthrows the legitimately elected government by force, the principle of democracy is violated and the dynamic balance of powers is disturbed. Initially, it seemed that democracy in Ukraine had a better chance of becoming a political device of the country than it did in Russia. However, this hope disappeared when the legally elected government was overthrown in Kyiv in 2014.

All and all, any revolution is a coup d'état, a rapid change of political poles. Faced with this rapid change, the whole body of the people cannot simultaneously roll over 180 degrees, and some parts of the body remain in the old position. Hence the Vendée rebellion in France 1793–1796, and the White movement in Russia 1917–1922. In Ukraine, a part of the people in the east of the country did not agree to accept the new, illegally established power, and a civil war broke out. All subsequent elections in Ukraine no longer included, in whole or in part, the eastern regions of

Ukraine. As a result, two opposite social narratives emerged: one in the east of the country (the Kiev regime is a tool of the West, guided by the ideology of superiority over Russia and carrying out genocide of Russians) and the other in the west of the country (LNR and DNR are Russian puppets, separatists, enemies of Ukraine). Since the roots of narratives are not in reason but in the living consciousness in the form of faith, logically narratives are indestructible, and only a compromise between them is possible, including a series of mutual concessions. The alternative is a war, but even the war is only a postponement of the compromise. These core social narratives, initially internal to Ukraine, quickly developed into TMNs between the West (which adopted the narrative of the Ukraine's new government) and Russia (which adopted the narrative of the DNR and LNR).

The fact that Western countries were acting under the influence of the TMN and not guided by the norms of international law follows from the fact that the new government of Ukraine, which came to power by forcibly overthrowing the country's legally elected president [6], was almost immediately recognized by the European Union and the United States, as well as the fact that over a period of almost 8 years of confrontation between the Kiev's regime and the LNR/DNR, the West did not help Ukraine resolve the conflict, for example, by facilitating the admission of the DNR and LNR into Ukraine on the basis of local self-government, as members of a federation. Countries such as the United States, Great Britain, Germany, Russia, and many others solved similar political problems by adopting a federal structure, and such a device in Ukraine suggested itself. Finally, the adoption of a federal structure by Ukraine was provided for by the Minsk agreements signed by Ukraine, Russia, Germany, and France in the Normandy Four format [7]. However, for almost 8 years, the Minsk agreements have not been implemented. Meanwhile, military clashes between Kiev's regime and the DNR and LNR, which the Western countries ignored, claimed lives of thousands of Ukrainians. The argument of the West for its indifference to the clashes was that the DNR and LNR are separatists and the West does not support separatism. However, this argument doesn't stand up to criticism, since disagreement to support the usurpers of power by the people of DNR and LNR is not the same as the will to separate from the country with legitimately elected government (e.g., the Catalonian separatist movement in Spain, or Scottish separatist movement in the

United Kingdom). And even in the countries with the legitimate government, under certain conditions people have the right to a referendum. For example, in 1992 Czechoslovakia peacefully broke up into the Czech Republic and Slovakia; similarly, in 2014 the people of Scotland had a legally recognized referendum on the right to secede from the United Kingdom, and only a small majority (55% of votes) in favour of remaining a part of the kingdom prevented the partition of the country. Thus, the reasons for ignoring the conflict in eastern Ukraine by the West rest not only in politics (e.g., the US interest in weakening Russia by hands of Ukrainians and strengthening its economic influence on the EU), but also in psychology, specifically in the TMN that Western politicians had adopted regarding Russia.

Why did Russia stop insisting on compliance with the Minsk agreements? The official grounds for the start of the military special operation were the suffering of the people of the LNR/DNR from systematic shelling, the build-up of armaments by Ukraine with the help of NATO countries, the prospect of Ukraine joining NATO and the threat of Ukraine renouncing the status of a country with no nuclear weapons, while the internal grounds were the fear of the possibility of a sudden attack by Ukraine on the LNR/DNR and Crimea and the loss of Sebastopol—the stronghold of the Russia's Black Sea Fleet. In addition, there has been a strong sentiment in most Russian people that Crimea, which till 1954 used to be a part of Russian Federation, was written down to Ukraine on the condition that Russia and Ukraine were one country, and when Ukraine separated from Russia in 1991 it lost the right to this territory populated predominantly by Russians. The importance of these reasons for Russia outweighed the understanding that the special military operation would lead to international isolation, but most importantly, in the eyes of the West, Russia would find itself in a morally flawed position of an aggressor. This is what happened. The confrontation between local social narratives of Russia versus Ukraine turned into an open confrontation between Russia versus the West TMNs, which completely deactivated the institute of diplomacy that used to regulate and balance differences in political views.

9.5 Russia and the West: Media Totalitarianism

At present, the totalitarian media narrative has developed in the West, in which Russia is depicted as an aggressor and invader. At the beginning of the special military operation, the author of this book was in the United Kingdom and personally observed the dynamics of the anti-Russian TMN in this country. Thus, initially British leader Boris Johnson objected in parliament to the ban on the Russia Today channel, which was demanded by the Labourites, arguing that we live in a democratic society and respect the freedom of speech, but gradually he lost ground under the pressure of the social narrative, and about a week after the start the special military operation in Ukraine the RT channel was blocked. Any statements in the British press that at least to some extent represented the arguments of Russia became accompanied by the epithets 'false', 'untrue', or 'erroneous'. Interviews with supporters of the Ukrainian narrative were lavishly published, whereas interviews with supporters of the Russian narrative were absent. The slogan 'Glory to Ukraine' began to be repeated with emphasis by the leaders of Western countries, and Russia's arguments regarding the tolerant attitude of official Ukraine towards the ideology of extreme anti-Russian nationalism (for example, the recognition of Stepan Bandera who collaborated with Nazi Germany as a hero of Ukraine, the statements of some political figures in Ukraine calling for the destruction of Russia and Russians, the education of Ukrainian youth in the spirit of denying everything Russian, clothes with Nazi symbols among the militants of some paramilitary formations, and demolition of monuments that commemorated achievements of the Russian empire or the Soviet Union) were ignored. Western countries have increasingly begun to implement packages of sanctions against Russia which hit not only the Russian elite but the entire Russian population. Thus, Russian airlines were initially banned in the United Kingdom, and a few days later in other EU countries. In other words, the anti-Russian TMN has, in part or completely, suppressed the principles of democracy in Western countries.

In return, in Russia a counter TMN in regard to the West emerged, depicting the united unfriendly West as a space of total lies, unlawful appropriation of the Russian state funds, double standards, hypocrisy, perverse morality, oblivion of traditional Christian values, and the desire to fragment Russia. Any objections to this narrative, and even simply its corrections, are perceived in most of Russian media as a betrayal of Russia's national interests.

On 7 October 2023 another ancient smouldering conflict of TMN's blazed in the Middle East, with the Palestinian paramilitary movement Hamas attacking Israel by killing over 1200 people and taking hostages, and the Israeli army retaliating by bombing Gaza and killing thousands of civilians. Like in the conflicting TMNs of Russia and the West, in the Palestinian-Israeli conflict the local social narratives, depicting negative images of their opponents, have been adopted by the wider parts of the globe, with some of the Western states taking side of Israel and many Muslim countries being on the side of Palestine. On the face of it we see another explosion of powerful irrational forces of social nature grinding up individual human lives, which in reality is a predictable outcome of the negative psychic energy of collective unconscious living consciousness that have been accumulating in these nations for millennia.

9.6 Conclusion and Cautious Forecast

Since narratives are charged with psychic energy by faith and emotions, they are not subject to logical arguments and change slowly, mainly under the influence of real events. As the resources of the parties in the military confrontation in Ukraine will be coming to an end, the attractiveness of resolving the conflict through negotiations will grow. At present, resumption of the negotiations that began in Istanbul on March 29, 2022, is unlikely. However, there is no alternative to a peace treaty, so gradually the negotiations should lead to a certain agreement and an end of the military confrontation. This would reduce the charge of psychic energy in both TMNs and defuse the tension between Russia and the West. Nevertheless, because social narratives have their roots in the depth of collective living consciousness, the victory of one of the parties is never final and the narratives may keep living for decades until tectonic forces of history replace the conflicting social narratives with new ones.

Since a social narrative is a phenomenon of collective living consciousness, it exposes the fact that in some fundamental features collective living consciousness differs from the living consciousness of an individual. In the next chapter, we will analyse these differences in more detail, and consider the ways of maintaining a balance in relations between individual and collective living consciousness.

References

1 Subbotsky, E. V. (2013). Russia in a conjured dream. *Psychologist, 10*, 40–83. https://e-notabene.ru/psp/article_11353.html
2 https://www.merriam-webster.com/dictionary/democracy
3 https://en.diagram.com.ua/aphorisms/aphorisms78.shtml
4 Afonasin, E. (2020). Protagoras of Abdera. Fragments and Testimonia. Institute of Philosophy and Law of SB RAS. https://cyberleninka.ru/article/n/protagor-iz-abder-fragmenty-i-svidetelstva-1/viewer
5 Kant, I. (1993 [1785]). *Groundwork of the metaphysics of morals* (J. W. Ellington, Trans., 3rd ed.). Hackett. ISBN 0-87220-166-X
6 https://en.wikipedia.org/wiki/2010_Ukrainian_presidential_election#cite_note-cec-declaration-1
7 Full text of the Minsk agreements. https://www.ft.com/content/21b8f98e-b2a5-11e4-b234-00144feab7de

CHAPTER 10

Golem: Living Consciousness Without the Brain

10.1 Individual and Society

Unlike external objectified consciousness that 'lives' on inanimate carriers (spoken and written languages, symbolic structures, and other human artifacts), the living consciousness is unthinkable without its connection to a living brain. Or is it? Can living consciousness function without the brain? Surprisingly, it can. Some called such living consciousness 'collective unconscious' [1], others 'crowd psychology' [2], still others 'swarm intelligence' [3]. In other words, living consciousness without the brain is the living consciousness not of an individual, but of a society. The problem that still warrants investigation is the relationships between the living consciousness of an individual and that of a society.

In this chapter we will discuss the most basic level of the relationships between consciousness of an individual and society, so basic that in certain respects these relationships are similar in human and non-human communities. To forestall accusations of biologizing and reductionism, I should say that more complex levels of the relations between living consciousness of an individual and society, which include science, art, and religion, are left out of the scope of this analysis. The reason is that it is at this most basic level of the relationships that the individual makes most important decisions about society and also makes the most tragic mistakes, while the more advanced levels of the relationships between individual and collective living consciousness are derived from these decisions and mistakes.

E. Subbotsky, *The Magic of Living Consciousness*,
https://doi.org/10.1007/978-3-031-54708-9_10

The aim of this chapter is to show that, although *living consciousness* of both an individual and a society are self-centred, there is a fundamental difference between the individual's and the society's *objectified consciousness* in that golem's living consciousness, when objectified in relations to individuals and other golems, stays immune to the imperatives of individual's morality. The difference arises from the fact that an individual cannot live without a society, but the society can live without the concrete individual and without other societies. Despite a society doesn't have brain of its own, it can nevertheless be represented as some kind of 'subject'. In this chapter, we will refer to this 'subject' using a metaphor of 'golem'.

10.2 The Concept of Golem

According to mythology, a golem is an anthropomorphic creature created from inanimate matter, such as clay or mud [4]. As stated by other sources, this humanoid creature can be magically animated. An individual can communicate with the golem and use it for his or her own purposes, however, the golem can exceed its 'authorities', get out of control of its creator, and do things that are indecent or even criminal for a human individual. In culture, mythology, and art, the concept of a golem varies. Thus, in the Polish author Stanisław Lem's novel 'The Golem XIV', a golem is presented as artificial intelligence to which "all motives of human thinking and behaviour are alien" [5, p. 23]. In this chapter, the term golem will be used as a metaphor denoting a community of individuals living together, united by common history, language, biological and psychological properties and possessing the supra-individual living consciousness. In wild nature, an example of such supra-individual living consciousness could be the so-called 'swarm intelligence'. Swarm intelligence exhibits itself in the coordinated purposeful behaviour of a group of insects, birds, or mammals. Swarm behaviour is an emergent behaviour arising from simple rules that individuals follow when they come together in large numbers. This behaviour is not exhibited by individuals and is not associated with any central coordination. For example, individual ants do not exhibit complex behaviour, however, an ant colony can perform complex tasks such as building nests, caring for their young, making bridges, and foraging [6]. At the level of human society, golems are peoples and countries.

The fundamental difference between a golem and an individual is their durability. Conceptually, an individual (lat. Individuum) is indivisible, which explains its fragility. A wound or a disease could result in death of the individual. In contrast, a golem can divide itself. Thus, during the swarming of bees, one colony can split into two or more separate colonies. Human golems can also divide themselves. Thus, in 395 AD, the united Roman Empire split into Western and Eastern ones, soon after which the Eastern Empire turned into an independent state—Byzantium. Byzantium lasted until 1453 AD, when the capital of the Byzantine Empire, Constantinople, was conquered by the army of the Ottoman Sultan Mehmed II. However, death of a golem does not mean the destruction of all its individuals. Thus, the Sultan offered the defenders of Constantinople to accept Islam and serve in his army, and some agreed, while those who disagreed were expelled from the city.

The longevity of a golem does not mean its immortality. Animal species exist for millions of years, but 99% of the species that existed on Earth are extinct by now. Ancient Egypt existed for more than 4 millennia before it became a province of Ancient Rome in 30 BC, and Ancient Rome existed for more than 1200 years. But unlike individuals, golems do not disappear entirely from the face of the planet: new animal species arise from pre-existing ones through speciation, and new empires appear in place of disappeared empires, which include individuals from previous empires.

Nevertheless, against the background of an individual's lifetime and its vulnerability to negative external factors such as illness or accident, a golem is surprisingly durable and resilient. Thus, because of the two world wars in the twentieth century some human golems lost millions of individuals, but not a single golem disappeared from the face of the planet. For example, Nazi Germany lost more than 8 million people in World War II, and the Soviet Union more than 26 million. Nevertheless, with the help of the United States Germany managed to restore its economy in a relatively short period from 1948 to 1960, and the Soviet Union, by and large, restored its economy during the 4th five-year plan of 1946–1950.

10.3 Golem's Consciousness

As stated above, a golem has a kind of consciousness. In animals, such a supra-individual consciousness is the aforementioned swarm intelligence. Going even further, some scientists consider the genetic code written on the DNA molecule as having some kind of 'consciousness'. The British evolutionary biologist Richard Dawkins writes: "We are survival machines – robot vehicles blindly programmed to preserve the selfish molecules known as genes" [7, p. 29]. In human societies, the consciousness of a golem, like that of an individual, can include objectified consciousness, in the form of myths and customs, and living consciousness, in the form of social narratives and 'crowd psychology'. Individual's living consciousness could dissolve in the crowd's living consciousness, and under the influence of the crowd an individual could behave contrary to his or her personal beliefs and values [8].

The living consciousness of a golem, like that of an individual, is fundamentally oriented towards the pursuit of its own interests, towards self-preservation and self-affirmation. However, there is a fundamental difference between these two kinds of living consciousness when they must convert into their objectified forms. An individual can exist only in a society, in communication with other people. The social nature of the individual makes the individual conform to moral norms that regulate relations between people based on the idea of justice. The basis of the individuals' morality is the principle of equality: although in their innate and acquired qualities individuals are not equal, the individual is obliged to recognize another individual as equal to himself or herself in relation to several fundamental principles, such as the right to life, personal property and the inviolability of the person. Moral norms are the objective framework that directs and puts limits on the individual's living consciousness, thus restraining the individual's egotism and fitting the individual's desire of self-realization into the frame of interests of other individuals.

On the contrary, a golem can exist independently of and alongside other golems, both psychologically and economically. Thus, some communities on the islands of Oceania or in the Amazon jungle have lived for centuries in isolation from modern industrial cultures. At the same time, in the communication between golems, both a struggle and a coalition with each other can happen. Thus, in animal communities, the struggle for food sources and territory prevails, but a coalition in the form of symbiosis can also take place. Peoples and states can cooperate and enter into alliances and treaties, but they can also fight for resources and territories. However, both the struggle and the coalitions of golems are based on the principle of their self-interests. One golem is not obligated to fulfil promises made to another golem if keeping the obligation becomes unprofitable to it. In other words, Immanuel Kant's categorical imperative "Do so that the maxim of your will may become the basis of universal legislation" does not work for the communication between golems. If in the relationship of individuals an individual violates the principles of law and morality there are bodies (i.e., public opinion, police, and court) that can pinpoint and correct the individual's action by punishing him or her. In the relationship between golems for a long period of history such organs did not exist. Therefore, conflicts between golems were resolved through wars.

Thus, most of the territories of modern golems are the result of military clashes that ended in favour of a more militarily and economically stronger golem. Nevertheless, in recent history there have been attempts to create over-state bodies with the goal of limiting the self-realization of golems. For example, because of the grim consequences of the First World War, during the Versailles Agreement in 1919–1920 the League of Nations was founded, which in certain periods of its functioning included 58 states. The goals of the League of Nations included limiting the self-centredness of the golems: disarmament, prevention of hostilities, ensuring collective security, settling disputes between countries through diplomatic negotiations. However, this organization could not effectively prevent self-centeredness of such powerful golems as Germany and Japan; the Second World War broke out, and in 1946 the League of Nations ceased to exist. After the Second World War, the countries participating in the anti-Hitler coalition created the United Nations (U.N.), which replaced the ineffective League of Nations. The purpose of the U.N. is to prevent wars. Since 2011, almost all sovereign states of the world—193 countries—have been members of the U.N. Nevertheless, according to the List of Interstate Wars, more than 70 military conflicts broke out since 1945, and the new wars are emerging. Even those golems that did not participate in military conflicts base their behaviour towards other golems on the principle of their own benefit. This means that the restraining influence of the U.N. is limited, and golems are now, as they have been earlier in history, not limited in their actions by the principles of individual morality, such as 'Thou shalt not kill', 'Thou shalt not lie', 'Thou shalt not bear false witness', 'Thou shalt not steal', 'Thou shalt not covet thy neighbour's house'. Indeed, during wars, killing combatants of an opposing golem is not considered a crime and is rewarded by a golem, while deception of the enemy, double standards in communication and even politically motivated killings are considered admissible.

As already mentioned, communication between individuals is based on the principles of individual morality (an exception is criminals—a special kind of individuals, the consideration of which is beyond the scope of this chapter), whereas golems communicate on the basis of commonality or differences of interests. More powerful golems absorb weaker golems or subjugate them economically by granting loans on certain economic and political terms. For example, an Ancient Roman golem destroyed a rival Carthaginian golem and subdued an Ancient Egyptian one. If a powerful golem fails to subdue a rival golem, it acts according to the ancient Roman principle of 'divide and rule'. Thus, during the Cold War, the Soviet super-golem supported golems that came out of the zone of influence of the competing American super-golem (such as Cuba or Nicaragua), and the American super-golem supported the golems opposing the Soviet one (Afghan Mujahideen during the period of Soviet expansion into this country). The same principle was practised by powerful super-golems regarding Korea (North and South Korea), Vietnam (North and South Vietnam), and other golems. After the collapse of the Soviet super-golem, the American super-golem included in its zone of influence the golems that had fallen away from Russia (the former countries of Eastern Europe, Georgia, Ukraine). Currently, there is a military clash between the Russian and Ukrainian golems. The world's dominant American super-golem cultivates this clash, supplying the Ukrainian golem in doses with weapons and political support, and receiving from this clash political (under the slogan of 'fight for democracy' in the former Soviet space), economic (strengthening its economic influence on the EU), and military (expansion and modernization of the military-industrial complex, testing military equipment and strategic potential) benefits, and the Russian golem plays on the interracial and inter-parties contradictions of the American super-golem. Competing with the American super-golem, the Chinese super-golem waits, acting on the ancient principle 'to sit on top of a mountain and watch two tigers fighting in a valley'.

In relation to its own individuals, a golem is as centred on its own interests as in relation to other golems, but since the existence of the golem is based on the existence of its constituent individuals, the golem has to put limits to its self-centeredness. If there is no external threat to the golem, then it is in the golem's interests to ensure the maximum prosperity of its individuals, and even to attract useful individuals from other golems (e.g., professionals or celebrities). If there is an external threat though, the golem is forced and able to sacrifice the well-being and even life of a part of its individuals, if this is necessary for its preservation.

The contradiction between objectified consciousness of an individual and that of the golem creates a psychological tension between the individual and the golem. Despite individuals' private living consciousness is free from moral values, when the individuals come into communication with each other their living consciousness becomes objectified and bound by moral laws (see Chapter 1). This, however, doesn't happen with the living consciousness of a golem, whose objectified consciousness, in the form of treaties and conventions with other golems, remains self-centred and free of moral obligations. It is psychologically hard for individuals with moral integrity to accept that their golem, with which the individuals are longing to identify themselves, doesn't conform to values of individual morality. Sensing this, the golem creates the way of acquitting its behaviour by making it appear conforming to the principles of individual morality.

Accordingly, there is a contradiction between the real actions of golems and their presenting these actions at the verbal diplomatic level: if at the level of real behaviour golems follow the principle of self-centeredness, in their verbal statements they turn to the principles of individual morality, by creating 'moral fakes' (see Chapter 8).

The organs of the golem, which carry out adjusting the golem's interests to interests consistent with the principles of individual morality, are sub-golems, or ideologists.

10.4 Faking Morality: Sub-golems as Ideologists

Metaphorically speaking, the body of a golem, like that of an individual, consists of organs. So, in an anthill there are casts of females, males, and workers, as well as ants of various specializations: builders, nannies, foragers, doctors, soldiers, and storekeepers. In a human golem, political parties, religious groups, and elite associations perform the role of organs. Some golem's organs (such as major political parties) are themselves micro or sub-golems, in relation to which the mother golem is a super-golem. In the elections, those parties come forward that at the moment maximally express the political and economic interests of the super-golem, though this is not always the case.

Sub-golems are much smaller than super-golems and less durable, so their objectified consciousness is closer to the individual's objectified consciousness and more sensitive to psychology and morals of individuals than is objectified consciousness of a mother super-golem. For example, sub-golems do not fight each other, but compete through opinion polls, public debates, and elections. There is also an agreement between sub-golems to conform to the Kantian categorical imperative, which forbids the sub-golem to seize the state power by force (see Chapter 9). The task of sub-golems is to reconcile the self-centred interests of their super-golem with psychology and morality of an individual. This is done by adjusting the super-golem's behaviour to the imperatives of individual morality, for example, by creating fake narratives, which interpret the super-golem's military actions or its failure to fulfil its promises as the actions that are morally justified, while accusing the hostile golems in lying and double standards. Such adjustments of super-golem's actions to imperatives of individual morality is called ideology, and the sub-golems and their leaders are ideologists.

10.5 Golem's Dreams

A dream is a form of living consciousness of an individual, in which the individual is completely immersed for a certain period of time, but on returning to the waking state considers the dream to be an inferior reality, or an illusion. Like an individual, a golem can have dreams in the form of collective narratives that guide the golem's actions for some time but ultimately lead to unacceptable consequences and are eventually rejected by the golem as illusions. Examples are the idea of superiority of the Aryan race over other races, which was the theoretical basis of the policy of Nazi Germany in 1933–1945, or the idea of building communism in the USSR in 20 years. The first dream was dispelled in 1945 because of the defeat of the German super-golem in World War II, the second collapsed when the leader of the Soviet super-golem Nikita Sergeevich Khrushchev was removed from leadership positions in 1964.

10.6 A Leader of Golem

A leader of a golem is a chosen individual who expresses the interests of the golem to the maximum degree and on whose actions the well-being of the golem depends. So, in the bee golems, the leader is a queen bee. The queen bee is not only 'responsible' for maintaining the golem's numbers, but also takes care of its structure: the glands of the queen's uterus secrete a special substance that it spreads over its body, then allowing the workers to lick it off, which suppresses the ability of the workers to lay eggs. In human golems, the leader is a chief of a tribe or the head of state. The leader's time of reign varies across golems. For example, the queen bee, with a decrease in the ability to produce offspring, is expelled from the hive or destroyed by the bees.

In human golems, a king or a queen passes on his or her power by inheritance, and a president is elected for a certain time. The way a golem chooses its leader is specific to that golem's living consciousness. Thus, a democratic form of choosing a leader is characteristic of some golems, and not characteristic of others (see Chapter 9 for more on that). The democratic form cannot be forced on a golem. Usually, after an overthrow of a monarchical rule authoritarian leaders come to power (Napoleon in France, Joseph Stalin in Russia, Muammar Gaddafi in Libya), whose rule differs little from the rule of the overthrown monarchs, and the overthrow of such leaders by the golem's internal forces with support of other golems leads to huge casualties among the population and bloody civil wars.

The leader of a golem does not rule alone but communicates with the golem through an elite group (government), therefore an attitude of an individual towards his or her golem, both positive and negative, often concentrates on the leader's personality and on the elite group. As a result, both the cult of the leader and attempts to forcefully overthrow or destroy him or her may arise. It also happens that with a positive attitude towards the leader, the attitude of the individual towards the golem's elite group is negative. If a leader of the golem, due to personal or objective circumstances, ceases to express the interests of the golem, he or she sooner or later loses power (palace coups in monarchies, impeachments or re-elections of the president in democracies, military or civil coups in authoritarian regimes).

10.7 Golem's Decay

Some golems (such as empires) are heterogeneous and consist of the remnants of other golems, historically united together. If the centrifugal forces of a mother golem (economic and cultural differences, conflicts of national elites, revolutions, and other) exceed centripetal forces, the mother golem may break up into smaller ones. Examples are the formation of independent 'republics' in Russia after the 1917 revolution, the separation of the Republic of Ireland from the United Kingdom in 1922, the collapse of the Soviet Union in 1991, the division of Czechoslovakia into the Czech Republic and Slovakia in 1993, the division of Yugoslavia into six independent states in 1991–2008. As a rule, the collapse is preceded, accompanied, or followed by diplomatic and military conflicts.

10.8 The Relation of an Individual to His or Her Golem

An individual may be dissatisfied with the actions of his or her or someone else's golem. This dissatisfaction can be caused by the inconsistency of actions of the golem with the rules of individual morality, insufficiently effective actions of the golem to protect its interests in relations with other golems, the inability of the golem to protect the interests of the individual, personal failures in the individual's life (failure in the profession, impossibility of self-realization, low earnings), and other reasons. However, this dissatisfaction cannot significantly affect the golem's living consciousness and actions. The first type of discontent is ineffective because the golem's living consciousness is insensitive to the rules of individual morality and is concerned exclusively with the golem's own interests; in addition, from the point of view of the golem's ideologists, the golem always has moral justifications for its actions. Other causes of the dissatisfaction do not affect the golem's living consciousness because the golem's living consciousness is cumbersome and has great inertia.

If individuals leave their golem, then the implicit reasons for leaving are usually hidden within the individuals' living consciousness, in the form of personal, economic and psychological dissatisfaction, although explicitly such individuals could explain their departure, both to others and to themselves, by the fact that the actions of the golem do not correspond to the rules of individual morality. By moving to another golem, such individuals may gain personal benefits, but they do not end up in a golem whose living and objectified consciousness is guided by the principles of individual morality, since the new golem is as centred on its own interests in relation to other golems and its own individuals as the abandoned one. Nevertheless, it may seem to the individuals that in the new golem there are principles of justice in relation to individuals, although in fact the new golem, due to the absence of an external threat or more plentiful resources, limits its self-centeredness in relation to its individuals to a greater extent than the native golem of the individuals.

If an individual unconditionally approves the actions of his or her golem, exhibiting the so-called blind or reactive patriotism, then such an individual must be ready to accept that the principles of individual morality are inapplicable to the golem's living and objectified consciousness and actions, although the individual may continue to apply these rules at the level of individual relations with members of his or her own or alien golems. In a more difficult position is an individual who exhibits civic or constructive patriotism, in which love to his or her golem is combined with disapproval of some of the golem's actions, and who calls to change these actions. Such an individual may join a sub-golem expressing his or her point of view (for example, a group of scientists who criticized the Soviet project of turning the Siberian rivers to the south), or individually oppose the ruling elite, while exposing himself or herself to the danger of being repressed (anti-fascists in Germany 1933–1945 years, dissidents in the Soviet Union). At the same time, an individual who professes constructive patriotism must be aware that the changes he or she advocates could lead to unpredictable consequences, sometimes opposite to those advocated by the civic patriot. Thus, the civic patriotism of the perestroika-era elites in the Soviet Union, represented by Mikhail Sergeevich Gorbachev and his group of supporters, who set positive goals for their golem, such as democratization of the management system and acceleration in the economy, led to the unpredictable collapse of the Soviet super-golem and the clashes between its former parts (for example, between Russian and Georgian golems, or between Russian and Ukrainian ones).

Another mistake of a civil (constructive) patriot may be projecting the norms of individual morality on relationship between golems. Thus, there is a statement that in the negotiations between the Soviet and American super-golems on the unification of Germany in 1990, an oral agreement was reached not to expand NATO into Eastern Europe, which was subsequently violated. However, from the point of view of the living and objectified consciousness of a golem, any agreement, oral or written, is respected only as long as it complies with interests of the golem and is violated in case of non-compliance, while the violation of the contract can be either recognized or not recognized by the violating golem. For example, the German super-golem unilaterally terminated the non-aggression pact between the German and Soviet super-golems signed in Moscow on August 23, 1939, starting Operation Barbarossa on June 22, 1941. On the contrary, NATO leaders argue that there was no agreement on NATO non-expansion to the East, since there are no written documents confirming such an agreement. However, even if such an agreement had been certified in writing and signed by the leaders of the NATO golems, it would have been violated as soon as the unification of Germany was completed and the contingent of Soviet troops withdrawn from Germany. An example is the second Minsk Agreement, adopted on February 11–12, 2015 at a summit in Minsk by the leaders of the German, French, Ukrainian, and Russian golems, signed by representatives of Ukraine, Russia, and the OSCE and later approved by the U.N. Security Council, which was not observed by the signatories of the agreement.

If the properties of golem's living and objectified consciousness had been taken into account by the ruling elite of the Soviet super-golem, then there shouldn't have been an agreement (oral or written) on the non-expansion of NATO to the East, and the condition for the unification of Germany on the part of the Soviet super-golem should have been termination of both military alliances—the Warsaw Pact and NATO. The ignorance of the specifics of golem's living and objectified consciousness by the patriotic elite of the Soviet super-golem in its negotiations with the American super-golem on the unification of Germany eventually led to the tragic confrontation between the former part of the Soviet super-golem (the Russian Federation) and the external super-golems (United States and EU), with unpredictable consequences for the whole of humanity.

10.9 Some Basic Rules that Make Sense for an Individual to Follow in Relation to His or Her Golem

Rule 1

Don't extend the imperatives of individual morality to your golem's actions. Remember that the golem always acts in its own interests (self-preservation and self-affirmation among other golems), and is ready to sacrifice some of its individuals, including you, to ensure these interests.

Rule 2

Even if you feel that your personal destiny has not turned out well, there is no reason not to respect the laws and rules of behaviour of your golem. Remember that the opportunity to practise your ethnicity (native language, mentality, customs, and habits), as well as to maintain moral behaviour towards other individuals, exists only because the golem protects you from external and internal enemies and dangers.

Rule 3

If you want to improve your golem while remaining inside it, do not try to act individually, as by acting alone you put yourself in danger. The golem's living consciousness is not sensitive to the actions of ordinary individuals and protects itself from such actions. Try to integrate into a sub-golem inside your super-golem (e.g., political party, initiative or religious group), which expresses your claims to your super-golem to the maximum extent.

Rule 4

Do not focus your dissatisfaction with the golem, if any, on the personality of the leader, since the leader expresses the interests of the golem's living consciousness and is not free in his or her actions.

Rule 5

When swapping your golem for someone else's, bear in mind that the new golem in its relations towards you and other golems is guided by its personal interests, and not by the rules of individual morality. Expect the rules of individual morality to work only in relations between individuals.

Rule 6

If you left your native golem, do not burn the bridges that give you the opportunity to return to it, even if you lived in other people's golems for years and decades. Life in someone else's golem may not be as attractive as it seemed at first, and your attitude towards your native golem may change. You could fit into someone else's golem, but you won't be able to completely shed off your ethnicity, such as the pleasure to think and speak in your native language or enjoy the poetry of your own golem to a greater extent than the poetry of someone else's golem.

Rule 7

During your lifetime, any sacrifice you can afford to make to your golem, other than sacrificing your own life, is justified by the fact that your golem's well-being means your personal well-being.

Rule 8

If a golem requires you to sacrifice your life for it, your decision will depend on your faith. Remember that it is not known for certain (and is impossible to find out) what will happen with your departure from the world and whether the world (and with it the golem) will be preserved. Sacrificing yourself for a golem only makes sense if you have faith that your golem will outlive you.

References

1 Jung, C. G. (1968). *Man and his symbols*. Dell Publishing Co.
2 Le Bon, G. (2004). The crowd: A study of the popular mind. Kessinger Pub. ISBN 0-7661-3008-8. OCLC 57245405
3 Beni, G., & Wang, J. (1993). Swarm intelligence in cellular robotic systems. In: *Robots and biological systems: Towards a new bionics* (pp. 703–712). Springer. https://doi.org/10.1007/978-3-642-58069-7_38
4 Idel, M. (1990). *Golem: Jewish magical and mystical traditions on the artificial anthropoid*. State University of New York Press. ISBN 0-7914-0160-X
5 Lem, S. (2020) Golem XIV. Act.
6 Maeterlinck, M. (1927). *The life of the white ant*. George Allen and Unwin
7 Dawkins, R. (1976). *The selfish gene*. Oxford University Press.
8 Toch, H. (1988). Psychology of crowds revisited. *Contemporary Psychology, 33*(11), 954–955. https://doi.org/10.1037/026204

The Mysteries of the Mundane: Epilogue and Summary

Going against the stream, Freud managed to claim irrationality for people's psyche, but his claim was conditional: he placed irrationality into the domain of unconscious. But when we look at the thoughts, feelings, and emotions that flow within our private conscious mind, we see that these subjective experiences ignore the laws of logic and morality. Despite our utmost desire to be rational and moral individuals, within our inner living consciousness irrational processes dominate. It is only when we are addressing others that we step on the solid ground of logic and morality and part with the magic of our private mind.

Being shy to acknowledge the ignorance of logic and morality by our living consciousness, we often prefer to ignore this part of the mind. Physicists explore atoms and vacuum, biologists discuss the origins of life, philosophers study concepts and phenomena via thought experiments, but the living consciousness employed in all these processes remains invisible. We could talk about the relationships between the concepts of brain and mind by conducting the 'brain in the barrel' mental experiment and still don't notice the invisible 'ghost' that builds these concepts and conducts this mental experiment. This invisible ghost is the subjective experience of our living consciousness. The most elusive part of the living consciousness is our internal Self—the agent which does the perceiving, thinking, creating, and remembering. When we try to catch this ghostly agent in the trap of our attention, we notice that we cannot do this: the agent escapes the focus of our attention and is always 'behind the head'. It

E. Subbotsky, *The Magic of Living Consciousness*,
https://doi.org/10.1007/978-3-031-54708-9

is easy to see the products of our internal Self imbedded in letters, drawings, and artifacts around us, but when we try to perceive the internal Self as such we enter the circle of the endless hunt for the 'homunculus'. When trapped into this hunt we become aware of the existence of living consciousness.

How then can we study the living consciousness, which can't be seen or perceived by senses? The simplest way is to use the 'fossils' of living consciousness. The source of fossils of living consciousness is the phenomena of the external world. We see shapes and colours, hear sounds, experience smell and taste. Comparing phenomena with each other, we create concepts, schemes, and theories and designate them with the help of symbols: letters, numbers, and words. These symbolic artifacts, or rational constructions (RCs) is what we call objectified 'fossils' of our living consciousness. RCs are created for nearly every aspect of living consciousness—subjective space and time, causality, and qualia (colors, tastes, smells, and other sensations). Still, the RCs are the external referents of living consciousness, but not the living consciousness itself. For addressing the living consciousness, we need a special image.

Speaking metaphorically, a suitable image for living consciousness could be that of a self-reflecting 'inner screen'. Like a movie that we see in a movie theatre, we observe our living consciousness on the inner screen of our mind. However, there is a fundamental difference between the screen in a movie theatre and the inner screen of our living consciousness: whereas the former displays the images that are projected on it from the outside, the latter is *reflecting itself*. If a movie character experiences pain, we have to infer his or her feeling by his or her behaviour, but if we experience pain, we don't need to make the inferences—we just feel the pain and know that we feel it.

The inner screen displays images of our body, images of people and objects in the outer world, and our feelings, emotions and thoughts. It also displays images not available in our perceptual field (dreams) or completely impossible in the real world. It includes our memories and plans, our thoughts about the laws of nature and morality. Finally, it includes our beliefs, religious or otherwise. In other words, the inner screen of our living consciousness is what is happening in our mind when we look 'inside ourselves'. But if the inner screen is so capacious, is there anything it does not include? Yes, there is. It does not include our internal Self—that very 'I' which is 'the watcher'. It also does not include our unconscious mind—all those processes that go on in that part of the iceberg that is hidden under the surface. It is only when the unconscious processes emerge on the inner screen that we become aware of their existence.

It is possible to assume that the inner screen isn't a prerogative of human living consciousness and that other living creatures have the inner screen as well. The presence of the inner screen is a feature that distinguishes living entity from a non-living one. The success of digital technology in the XXth century briefly created the hope that the door to the living consciousness was finally found and an objective 'model' of living consciousness in the face of advanced computers was created. But this hope didn't last. Soon it became clear that even the most advanced computers cannot have the inner screen of consciousness; like the images on the screen in the movie theatre, computers software are instructions programmed by a human person. These instructions can be complex and the computer's 'behaviour' might look indistinguishable from that of the living entity, but in the end the computer only regurgitates the experience of people converted into sophisticated algorithms. The computers lack the most essential part of living consciousness—the self-reflecting inner screen. By the same token the most complex molecules, such as RNA, which are vital for functioning of living creatures, don't have the inner screen either and operate on the programme laid down in them by nature.

Perhaps, the awareness of the principal limits of the AI wouldn't be enough to turn the attention of scientists towards the wonders of their living consciousness had it not been for the discovery of some enigmatic physical phenomena. It became increasingly evident that descending deep into the microworld is impossible without taking into consideration that it is our living consciousness that does observations and measurements. Researchers of the mega-world stumbled upon enigmatic phenomena as well: It turned out that the universal cosmological constants are 'finely tuned' to each other in such a way as if they had been arranged by someone to make the emergence of life and people possible. The controversial 'entanglement effect' in the quantum world added to the list of unresolved enigmas in physics. With all the success of biological sciences, the puzzle of the emergence of life from inanimate matter remains unresolved. Psychologists and philosophers struggle to explain how the wonderful worlds of colours, odours, and tastes emerge from the electrical impulses cycling within the brain. In medicine, the fantastic placebo effect is registered and used in practice, but still escapes scientific explanation.

Physicists and neuroscientists were not alone in their reanimated interest towards enigmatic nature and functions of living consciousness. Cognitive psychologists, who used to measure themselves on the scale of natural sciences by studying 'hardened' pieces of living consciousness in the form of mental habits, became more conscious about the fact that there are phenomena in our everyday life which escape scientific explanation. Disabled people who control their rolling chair by thinking alone—how can they make the inanimate physical object react to their nonphysical thought? Our night dreams—where do they come from? Some of them are simply a response to events in our everyday life, but others are unusual and mysterious. But the most magnificent and enigmatic is our everyday subjective experience: our thinking and imagining. Inside our minds, we can travel back in time or forward into the future, instantly get ourselves at the edge of the visible universe or inside a black hole, turn ourselves into a shape of a cat, fly in the sky like a bird or swim under water like a fish. Inside the mind, we can ignore logic and morality and fulfil our wildest desires without consequences.

In this book we tried to show that living consciousness we experience every day without even noticing it conforms to the laws that transcend the laws of formal logic. These laws are the laws of magic. Far from being a mere accompaniment to the brain functions, living consciousness defines the features of both physical and social universes. We attempted to examine how our inner mental processes (thinking, imagining, dreaming, and desiring) work in our everyday life, and how these processes could be observed, studied, and used for practical purposes in life. We discovered that our everyday living consciousness harbours phenomena inexplicable in terms of scientific knowledge—the real 'wonders of the mundane'. So, what are these phenomena?

The most mysterious of them is the phenomenon of participation, which, paradoxically, is also the most mundane event of the everyday life. Through participation, our nonphysical mind can control voluntary movements of our body and patterns of our EEG. Another typical example of participation is love. A person who was in love with another person knows that the object of love—willingly or not—acquires power over the person who is in love. In the person who is in love the instinct of self-preservation is suppressed, and the person can sacrifice for the loved one the most valuable parts of his or own life—career, honour, property, and sometimes life itself. Literary descriptions abound: Romeo and Julia committed suicides when they thought that their loved one was dead, the character of the novel by Stefan Zweig 'Amok' gave up his career and life with the aim of saving the reputation of a woman he had suddenly fell in love with, and the list of examples could go on. There is something in 'being in love' that can't be reduced to natural causes, and this something is the fact that the object of love, being a part of the external world, simultaneously is a part of our living consciousness. We are in love not with a real person 'out there', but with the image of that person we created in our mind. Love is a case of the participation between two elements within our living consciousness: our need to fall in love and the image of the person we chose to fall in love with.

But love towards another person isn't the only form of love. There is also love to God, to our profession, or just to a certain physical object. For the sake of having access to the object of love a person who is in love is ready to give up his or her most vital needs. Suicidal terrorists blow themselves with the words of their love to God, and collectors are ready to pay fortunes for getting their hands on the object of their passion. People can even fall in love with their financial dream, like the marvellous but naive Pinocchio who listened to the crooked cat and fox and planted his coins in the Miracle Meadow, hoping that the tree would soon grow dripping with gold coins. Participation with seductive financial offers is another everyday magical experience, which makes people vulnerable to magical manipulations of their living consciousness. Altogether, *the ability to experience participation* is the most prominent and mysterious everyday phenomenon that comes into the domain of 'wonders of the mundane' (Chapters 1, 2, 7, 8, and 9).

Subsequent analysis revealed that living consciousness transcends the boundaries set by formal logic, such as the laws of identity, contradiction, and excluded middle. In the magical world of living consciousness, a part can include the whole, two separate things can be one, and an effect can precede its cause. Without having a formal structure, living consciousness precedes and participates in shaping the advanced forms of consciousness, such as logical thinking, social and cultural forms of perception, and moral behaviour. In science, living consciousness in the form of creative imagination is the way of producing novelties literally 'from nothing'. *The magical ability of living consciousness to produce novelty by transcending the laws logic and causality* is a manifestation of the wonders of the mundane (Chapters 2 and 4).

Further, psychological studies of infants' cognitive abilities revealed that living consciousness is present from birth and even before birth.

While still in their mothers' womb infants can distinguish voices of their mothers from voices of strangers, newborn infants can discriminate shapes and magnitudes, and a few month old infants are capable of understanding some important properties of physical objects, such as causality and permanence. *The presence of living consciousness from birth* is the 'everyday wonder', which is hard to overstate (Chapters 2 and 3).

The wonders continue with older children, who like listening to fairy tales and watching cartoons with magical events before they learn to read. For more advanced ages there exist such masterpieces of literary magic as fairy tales by Brothers Grimm and Hans Christian Anderson, 'The Chronicles of Narnia' by C.S. Lewis, and many others. Children's games of pretend abound with magical events. In play, a Lego block can turn into a piece of cheese and a wooden stick into a running horse. When engaged in games of pretend, children can fight monsters and dragons, speak with trees, and fly in the air. The onset of gadgets in the twenty-first century brought the magical world of images in a nursery. The studies showed that children's magical thinking is beneficial for their cognitive development. This makes the *children magical thinking* another candidate for the wonders of the mundane (Chapter 3).

The psychological universe of our living consciousness is not the only domain that hosts the everyday magic. Surprisingly, the physical universe harbours magical events too, if you know where to look. Thus, classical physics forbids the 'mind over matter' magic, such as affecting a result of throwing a coin by just thinking hard about a desired outcome. However, the emergence of quantum mechanics in the twentieth century shattered this prohibition, thus opening a gate for the living consciousness to directly affect physical reality. It turned out that the observer's living consciousness, through its objectified extensions—measuring devices—plays a definitive role in what the observed quantum reality is. The *participation of observer's living consciousness in determining certain quantum processes* can be added to the list of the wonders of the mundane (Chapter 4).

Participation of living consciousness in shaping some elements of physical universe is mysterious enough, but the existence of living consciousness is a mystery of its own. How come that a person, who in terms of classical physics is an insignificantly small part of the physical universe, has the whole universe within his or her mind? In the Personal Universe of our living consciousness, contrary to common logic, a part can contain the whole. The *presence of the whole physical universe within our individual living consciousness* is another mystery of the mundane (Chapter 4).

Further, because our living consciousness is finite in time and we sometimes temporarily lose it, we must assume that it was created by someone beyond it. We know that our bodies emerged within physical universe, but how did our living consciousness come to be? All of the attempts to derive our living consciousness from the physical universe (e.g., through evolution of the brain) run into the logical circle, since the physical universe itself doesn't exist without living consciousness creating the image of this universe. This makes one assume the existence of some kind of Creator, which some people call God and others call the Supreme Programmer. But this assumption creates another mystery: We have to accept the assumption that the Creator we produced by our imagination must exist outside our living consciousness. The reviewed studies showed that both children and adults view such an assumption impossible, and adults succumb to this assumption only when prompted by observing another mystery—the magical conversion of their mental image into a physical object. Thus, we can add another wonder that defies logic to the list of 'wonders of the mundane'—*the Supreme Programmer we created within our living consciousness must exist outside our living consciousness* (Chapter 5).

Let's continue by asking how, despite our individual life is short, are we able to extend our living consciousness back in the past till the beginning of the universe and forth into the future till the end of the universe? In making these extensions, we use the knowledge and power of our faith to bring the events that we don't see into the present time, as if these events were happening now. Our *ability to turn stories about the past into reality and project our life into the unobservable future* is another mystery of our living consciousness and should be placed in the 'collection box' of the wonders of the mundane (Chapter 6).

The analysis of changes in our understanding of how our living consciousness functions brought us to the view that our living consciousness not only reflects reality but also participates in creating it. Whereas people of the past centuries believed in that nature and society are ruled by God, today most of us believe that our living consciousness can influence reality. The *belief that our living consciousness participates in creating reality* is another wonder of the mundane (Chapter 6).

The fundamental property of living consciousness to participate with suggested messages is a double blessing. On one hand, it allows us to emphasize with people in trouble and maintain principles of morality, but on the other hand it makes us vulnerable to suggestions and manipulation of our living consciousness. The *ability of a linguistic message to reach our living consciousness and make us conform to messages that contradict our rational judgement* is a psychological mystery that deserves to be listed among the wonders of the mundane (Chapter 7).

The tendency of our living consciousness to participate with suggested messages opens the opportunity to deceive us by creating false duplicates of things. As a result, the reality acquires a 'doppelganger'—the fake-reality. Living in the world of fake-reality, we often struggle to tell real objects and events from fakes. The production of fake-reality makes our modern life complicated and opens the gate to deception and fraud, but it can also be used to solve problems in communication, via producing 'useful fakes' in the form of politeness, courtesy, and diplomacy. The *ghostly world of fake-reality* puts one more coin into the 'collection box' of wonders of the mundane (Chapter 8).

The fake-reality expands into social narratives—the 'urban myths' that circulate within the living consciousness of a society. The most powerful of the social narratives are totalitarian media narratives (TMN), particularly in their political manifestation. These narratives rooted in history have a hypnotizing power over the living consciousness of individuals. The most tragic manifestation of TMN is the conflict between modern Russia and the united West, which has its physical embodiment in the war in Ukraine. The *magical power of TMN over living consciousness of individuals* is another phenomenon that escapes the claws of logic and rationality, thus adding to the wonders of the mundane (Chapter 9).

Last but not least, the individual's living consciousness, which conceptually is linked to a functioning brain, finds its analogy in the living consciousness of a society. The most fundamental difference between the living consciousness of an individual and that of a society is that the latter functions without a link to the brain. The *'brain free' living consciousness of a society* adds to the list of wonders of the mundane (Chapter 10).

The list of mysteries among which we live presented in this book is far from being complete, and even in many of those presented we only scratched the surface. But it might be enough to demonstrate that our daily existence is full of invisible things that cannot be explained by science and formal logic—the wonders of the mundane.

Index

E. Subbotsky, *The Magic of Living Consciousness*,
https://doi.org/10.1007/978-3-031-54708-9

D

E

The manufacturer's authorised representative in the EU is Springer Nature Customer Service Centre GmbH, Europaplatz 3, 69115 Heidelberg, Germany. If you have any concerns regarding our products, please contact ProductSafety@springernature.com

Printed and bound by CPI Group (UK) Ltd, Croydon, CR0 4YY
01/10/2026
02225903-0002